ANALYSIS
AND CONTROL
OF PRODUCTION SYSTEMS

PRENTICE-HALL INTERNATIONAL SERIES
IN INDUSTRIAL AND SYSTEMS ENGINEERING

W. J. Fabrycky and J. H. Mize, Editors

ANALYSIS
AND CONTROL
OF PRODUCTION SYSTEMS

Elsayed A. Elsayed
Thomas O. Boucher

Rutgers University

PRENTICE-HALL, INC., Englewood Cliffs, New Jersey 07632

Library of Congress Cataloging in Publication Data

ELSAYED, ELSAYED A.
   Analysis and control of production systems.

   (Prentice-Hall international series in industrial and
systems engineering)
   Includes index.
   1. Industrial engineering.   2. Production control.
I. Boucher, Thomas O.   II. Title.   III. Series.
T56.E56 1985   658.5   84-6824
ISBN 0-13-032897-9

Editorial/production supervision
and interior design: Theresa A. Soler
Manufacturing buyer: Anthony Caruso

Printed in the United States of America

10   9   8   7   6   5   4   3

ISBN 0-13-032897-9 01

Prentice-Hall International, Inc., *London*
Prentice-Hall of Australia Pty. Limited, *Sydney*
Editora Prentice-Hall do Brasil, Ltda., *Rio de Janeiro*
Prentice-Hall Canada Inc., *Toronto*
Prentice-Hall of India Private Limited, *New Delhi*
Prentice-Hall of Japan, Inc., *Tokyo*
Prentice-Hall of Southeast Asia Pte. Ltd., *Singapore*
Whitehall Books Limited, *Wellington, New Zealand*

To
Linda, Aladdin, Amena, Amira, and Amardean

To
Unn

CONTENTS

PREFACE

The production of goods is one of the basic functions of any society. Each year in the United States about a quarter of the national income is generated by about 10% of the population working in manufacturing industries. Today, the rate of output per worker in those industries is about twice what it was 25 years ago and over five times what it was at the turn of the century.

The production engineer is one of the people responsible for operating these production systems and improving them in order to maintain this historical trend in productivity. One of the primary functions of the production engineer is to coordinate the use of machinery and workers and to make decisions as to what is to be produced, how much, and when. While making these decisions, production engineers need to analyze and control the operation of the production system such that its productivity is maximized and the manufacturing cost per unit produced is minimized.

This book is about the analysis and control of production systems. Each chapter focuses on one of the primary activities that compose the analysis and control function. After an introduction in Chapter 1, we describe the forecasting techniques that are used to estimate the demand that the system will have to respond to over time (Chapter 2). In Chapter 3, methods of controlling inventory levels consistent with these demand forecasts are discussed. Techniques for planning the aggregate activity level of the production system are discussed in Chapter 4. In Chapter 5 we present the subject of material requirements planning, which ties together both inventory decisions and production planning decisions and is especially relevant to small batch and discrete-parts manufacturing. Chapter 6 describes scheduling techniques that are used by production engineers in planning and executing projects, such as plant expansions or the installation of new equipment. Chapter 7 deals with the lowest activity in production control decision making, the sequencing and scheduling of jobs on machines. Finally, in Chapter 8, we describe the changing environment of discrete-parts production, particularly factory automation, and discuss the implication of these changes for analysis and control decisions.

Throughout the writing of this text we have tried to adhere to two principles. The first is to emphasize the *process* of production control decision making. This should be reflected in the order in which subjects are sequenced in the text, as well as our attempts to describe the connection between the activities which these chapters represent. The second principle is to relate the techniques being described to the kind of industrial environment (i.e., production system organization) in which they are most likely to be usefully applied. We hope we have accomplished this objective, particularly in the chapters on production planning, material requirements planning, and operations scheduling.

This book is written specifically for senior undergraduate and first-year graduate students in industrial, production, manufacturing, and systems engineering disciplines. The book presumes a background in differential calculus, mathematical programming, and statistical theory up to regression. It is our experience that all the material in this book can be covered in a one-semester course. We have provided a solutions manual for the problems at the end of the chapters.

Although we have included some results from our own research, a textbook does not represent the work of just the authors. This is reflected in the many references throughout the book; we hope we have given due credit to everyone concerned. Particular acknowledgment is made to Taylor & Francis, Ltd. for permission to use some material in sequencing the static flow shop which appears in Chapter 7, and to John Wiley & Sons, Inc., for the use of Figures 3-1 and 3-2.

We would like to thank the students of the Department of Industrial Engineering at Rutgers University, who had to live with and suffer through early versions of the manuscript. We are indebted to Mukesh Balani for his enthusiastic assistance in developing the solutions manual, redrawing all the figures for this manuscript, and making helpful recommendations along the way. Due thanks are extended to Doris Clark, who helped type revisions of the original manuscript and provided support in other aspects of completing the text.

We are very grateful to our former teachers and colleagues who have influenced our direction of research and our thinking about this subject. Professor Boucher would like to acknowledge a special debt of gratitude to William Maxwell of Cornell University, who, as a former colleague, was an important influence in encouraging his interest in the subject.

We reserve the greatest thanks for Linda Elsayed, who gave so freely of her time and worked so hard to make this book possible.

Elsayed A. Elsayed
Thomas O. Boucher

ANALYSIS
AND CONTROL
OF PRODUCTION SYSTEMS

INTRODUCTION

1.1. The Production System

This book is about the analysis of what is referred to as a *production system*. Loosely described the production system is the collection of material, labor, capital, and knowledge that goes into the manufacture of a product. How this collection of components is put together in any specific situation defines a particular *system*.

Although there are always differences between particular production systems, such systems can be classified by different criteria and meaningful analysis applied. For example, production systems can be categorized by industry: steel, auto, pharmaceutical, and so on. Such a classification is made when the main interest is the product or some general characteristics of the production technology.

Another taxonomy of production systems is to classify them by production flow characteristics. By virtue of the nature of a product and the demand for the product, some goods are made by production processes that are continuous, whereas other goods are best made in a discrete or one-at-a-time fashion. Products in the former group are petroleum, pharmaceuticals, and chemicals; the latter group contains ships and special-purpose machinery. Between those two extremes, there are products made in small lot sizes, mass-produced goods, and in some chemical and food product industries, batch (as distinct from continuous) process production.

1

The methods that will be discussed in this book are often tailored to different classes of production systems. Where appropriate, a discussion of these methods will be accompanied by a description of the production environment.

1.2. Production System Analysis

There are three major building blocks in a production system. These building blocks are:

1. The machine technology base, or tools of production
2. The organization of the production system
3. The techniques of production management that are applied to control the operation of the system

It is the third area that we focus on in this book. Each of these building blocks is important and warrants some discussion.

The available machine technology sets the boundaries on the processes that can be employed in converting inputs to finished products. Decisions with respect to which machine technology to employ in a production system are in general, long-term planning decisions. Over the long term, they are among the most important decisions made by a manufacturing firm, since it has been shown that the major component of productivity improvement in American industry is directly traceable to the employment of new machine (production) technology. Since this book is concerned with short-term production control decisions, we assume that the available set of machine technologies is fixed. Information as to the characteristics of these existing machines is relevant to particular analysis and control decisions, and such information is part of the information base used in making such decisions. This will be described shortly.

The second major building block is the organization of production. One distinction often made in discriminating between production organizations is to classify production facilities by layout: product layout versus process layout. The *product layout* organizes people and machines in a manner satisfactory to the production of a single product (or limited range of similar products). Production machinery is set up for the performance of sequential operations on that particular product. Examples of this kind of production organization are container manufacturing plants, chemical plants, automotive assembly lines, and textile plants. Because a product layout usually includes several sequential operations, organized in a production line, it is often referred to as the *flow line* organization of production.

At the other extreme, manufacturing facilities can be organized according to the manufacturing processes involved in the production of products. In such an organization, machinery is grouped within the plant based on common purposes: for example, a grouping for lathe, milling machines, grinders, and so on. This is called a *process layout* and is usually found in situations where there are a large number of different products and components using the same machinery, but not necessarily in the same order. Examples are metalworking machine shops, general industrial machinery manufacture, farm machinery manufacture, and aircraft parts production. The term *job shop* is often used to describe a plant manufacturing a variety of items with different machinery requirements and routings using a process layout. From the point of view of dealing with product variety, the process layout is the most flexible production organization; however, it has the disadvantage of being the most difficult on which to impose controls.

An intermediate form of production organization that is relatively new is called *group technology*. Group technology is intermediate between flow line and job shop production configurations. In situations where one is producing numerous products in small batches requiring different operations, the job shop form of production organization naturally recommends itself. However, if a large proportion of the components produced utilize roughly the same machines and have the same or similar machine routings, that group may be separated out and manufactured with a flow line arrangement using machines that are dedicated just to the manufacture of that group of products. The benefit of doing this is to simplify the production control problem, since a flow line is inherently simpler to manage than a job shop. In the United States, there have been many successful applications of group technology for portions of plants and for entire manufacturing establishments over the past 15 years.

In this book we assume that the organization of production has been determined. We explicitly take account of this organization in many of the models we describe because these models are often specific to production systems organized as either a flow shop or as a job shop. Where it is necessary to make the distinction, the discussion will be accompanied by a description of the appropriate organization.

This text concentrates on the third building block of the production system: the *analysis and control of production systems*. For the machinery of production and the organization of production to be effective, there must be a decision-making apparatus which determines, during any period of time, what products will be produced, how much will be produced, and when they will be produced. Furthermore, provision must be made for the timely procurement of raw materials from suppliers to the production system and there must be coordination between operations and departments within the production system. The techniques and methods developed to accomplish these tasks comprise the chapters of this book.

1.3. The Information Base

Production control methods are applied in consideration of both the machine technology base and the organization of production. This is accomplished by using an information base that completely describes the product and its manufacturing process in terms of machine requirements and the flow of the product within the production system. Since we will be using such data in the examples presented in this text, the main information sources will be described at this time. Briefly, they are the product bill of materials, the production routing sheet, the operations process chart, and the standard time and standard cost data base.

The product *bill of materials* (BOM) is a master list of components, purchased parts, and subassemblies that are required to produce a complete product. This document provides the engineer with information on all the pieces that must be brought together in order to deliver one unit of the final product. The bill of materials is also the key to what raw materials will be required.

The BOM does not define the operations that are to be performed in order to manufacture the product; this is the function of the component *routing sheet*. The routing sheet describes the sequence of work centers through which a part will travel and provides information on the operations that will be performed at that work center. The routing sheet also provides time standards for each operation.

The bill of materials and routing sheets are tied together into a description of the workflow by what is called an *operations process chart*. The process chart illustrates the flow of components and their final assembly into the end product.

From the point of view of planning and scheduling, we have described the most important pieces of information required. Additional information that must be provided is a forecast of demand and knowledge of the capacity of the plant in each machining center. As a general rule, the objective of planning and control decision making is to minimize some function of production cost, often subject to other constraints. For this reason, adequate information concerning cost must be available through the plants' accounting system; for most firms this is accomplished through the standard cost data base. Throughout this text we employ the data of these information sources in the course of explaining production control methods and techniques.

1.4. Organization of This Book

This book is organized largely in accordance with the chronology of events that take place in arriving at production control decisions.

The production decision begins with estimates of customer demand; therefore, the first major topic in the book is *forecasting* (Chapter 2). Forecasting is useful primarily in companies which produce a product prior to holding a firm order for it: for example, automotive manufacture. These companies usually produce a standard product to finished-goods inventory; the product is drawn from inventory as shipments are made. Forecasting techniques are less useful in situations where firms manufacture custom products to order, such as specialized castings. In such cases the plant will operate largely off the backlog of existing customer orders.

Chapter 3 describes some important models of *inventory* theory. The inventory decision pervades most of the other topics in the book and should be covered early. The finished-goods inventory is the buffer between demand and production. When one makes a production decision based on a forecast of demand, one is implicitly making an inventory decision. The rate of production in combination with the rate of demand completely determines the level of finished goods inventory. Similarly, raw materials inventory decisions are dependent on production decisions and form a key link between suppliers' production and plant production. Models relevant to making optimal production decisions in light of these relationships will be examined.

When a forecast is completed, the next step in the planning process is to plan for the utilization of plant resources in meeting the forecast of end product requirements. There are two levels at which this problem is approached: they are the subjects of Chapters 4 and 5. Chapter 4 describes production planning models used at the *aggregate planning* level, and in particular, in conjunction with the *hierarchical* planning approach. In this approach one plans for the optimal combination of manpower requirements and inventory requirements in order to meet a given demand forecast. The output of this process is a description of resources utilized and an overall schedule for production (*master schedule*). The master schedule describes, by time period, which products will be produced. This schedule usually does not deal with the problem of timing the movement of parts through the manufacturing operations within the plant, nor does it deal with the coordination of completion times for final assembly. Consequently, these aggregate plans are subject to further refinement in order to be implemented in the production environment.

The subject of Chapter 5 is *material requirements planning*. MRP is an approach to disaggregating a master production schedule to meet final demand while considering the problem of the timing of product flows within the manufacturing facility. MRP is credited with reducing work-in-process inventory within the plant through the control of these intraplant flows. Due to its current popularity as a control strategy in discrete-parts production systems, an entire chapter is devoted to describing this unique approach to disaggregating master schedules. The MRP approach is currently being used in thousands of production facilities.

Chapter 6 covers an important topic related to production decisions: *project planning and scheduling*. An example of this would be the installation of major pieces of machinery or major additions to the manufacturing facility. These activities, if not planned well, will involve some disruption of normal production schedules. Methods of project scheduling under resource constraints are also presented.

The output of the production planning process is a description of the set of jobs to be processed by work centers during specified periods of time. This brings the problem of production control down to the individual work center or machine level. In Chapter 7 we describe methods of scheduling and sequencing jobs through machines and work centers on the production floor. This is referred to as *operations sequencing and scheduling* and is the lowest-level activity in the production control process.

Finally, in Chapter 8, we take a look at the future of production analysis and control by investigating some of the more recent developments in dealing with the production control problem. We conclude the chapter with a discussion of factory automation and the new issues it raises with respect to production control.

REFERENCES

[1] MIZE, JOE H., CHARLES R. WHITE, AND GEORGE H. BROOKS, *Operations Planning and Control.* Englewood Cliffs, N.J.: Prentice-Hall, Inc., 1971.

[2] TURNER, WAYNE C., JOE H. MIZE, AND KENNETH E. CASE, *Introduction to Industrial and Systems Engineering.* Englewood Cliffs, N.J.: Prentice-Hall, Inc., 1978.

FORECASTING
AND TIME-SERIES ANALYSIS

2.1. Introduction

The first step in planning the operation of a production system is determining an accurate forecast of the demand for the items to be produced. This forecast is then used as a basis to specify the control policies for the inventory system, to load the machines, to determine the machinery and materials handling requirements, and to determine the work-force level during production periods. Forecasting is not only used to estimate the demand for products but is also widely used in service and nonmanufacturing systems. For example, the Department of Transportation is interested in freight demand arising from the need to ship commodities during any period of time. As will be mentioned later, forecasting is a time-domain process; that is, forecasting models provide us with a forecasted value at a given interval of time. The accuracy of a forecast is dependent on the accuracy of the data, the stability of the data-generating process, the length of the forecasting period, and the forecasting method used. Randomness of the data makes accurate forecasting difficult, if not impossible, to achieve.

The purpose of this chapter is to present some of the forecasting techniques and their limitations. Although these techniques are widely used for both short- and long-range forecasting, their values diminish as the time horizon increases.

2

2.2. Forecasting

The idea behind any forecasting method is to use past data in order to predict (project) future values. One way of categorizing the techniques used in forecasting is as follows.

Qualitative Techniques (or Judgmental Forecasting): These techniques are utilized when no or very few historical data are available. In these methods, experts' opinions and their predictions are considered the ultimate forecasted values. In forming their opinions, experts usually refer to similar situations and analyze the limited data in order to reach forecasted values. This approach to forecasting is often referred to as the *Delphi method*. Historical analogy, market research, customer surveys, panel consensus, and SWAG (sophisticated wild arbitrary guessing) are also qualitative forecasting techniques that can be classified under judgmental forecasting. Although not commonly used, these qualitative techniques may be the only method available to forecast, for example, the sales of a new product.

Quantitative Techniques: In these techniques, the historical pattern of the data is used to extrapolate (forecast) into the future. There are two main quantitative techniques: (1) time-series analysis and (2) structural models. The former treats a sequence of observations as a function of past history. The latter is an important branch of economics today and is widely utilized as an analytical tool in forecasting economic behavior. In this chapter we emphasize the techniques of time-series analysis, such as moving averages and exponential smoothing methods. However, we will begin with a discussion of the technique of regression, which is a widely used tool to model both time-dependent and structural relationships.

A second way of classifying forecasts is according to the forecasting range. *Short-Range Forecasting:* This forecasting is more accurate than medium- or long-range forecasting. A typical period for short-range forecasting is from hours to 1 year. Therefore, hourly, daily, weekly, and monthly forecasts are considered short-range forecasting. For example, electric utility companies use hourly forecasting of kilowatt-hour demand, while the production planning of manufacturing systems is usually based on the monthly forecast of unit sales.

Medium-Range Forecasting: The time period of the medium range is from 1 to 5 years. The 1-year forecast is the most accurate, while the 5-year forecast is the least accurate; this is due to the increase of uncertainty in the stability of the underlying data generating process with the increase in the period of forecasting. A typical example of medium-range forecasting is the enrollment of students in colleges and universities.

Long-Range Forecasting: The time period of the long range is more than 5 years. It is a difficult task to forecast economic events for 5 or more years, due to variations in the characteristics of the underlying process (e.g., technology and the state of the economy will change over the long term). Hence long-range forecasting of economic variables often requires a prediction concerning the state of these independent variables. Freight demand for commodities is forecasted for more than 5 and less than 10 years. Telephone companies use long-term forecasting to determine the future demand for telephone service in order to allow for a reasonable and cost-effective expansion of the network.

2.3. Forecasting Procedures

The first step in determining which forecasting procedure should be used is to plot the historical data as a function of time. One might suggest that after plotting the data, an exact curve that passes through each of the points could be used for accurate forecasting. In contrast, it will be shown that the exact fit may, in many cases, yield very poor forecasted values. In the following section some exact curve-fitting methods and forecasting techniques are presented.

2.3.1. GREGORY–NEWTON INTERPOLATION FORMULAS

The Gregory–Newton interpolation formula is used for interpolation and also for extrapolation (forecasting) when the function $f(t)$ is known at discrete, evenly spaced points. The last point is used as the base line. It is known that if there are n equally spaced data points, there exists a polynomial of a degree less than or equal to $n - 1$ which passes through every data point. This polynomial can then be used to forecast future values. Now we consider the following two formulas using a tabulated function and a difference table, for interpolation or extrapolation.

Gregory–Newton Forward-Difference Formula:

$$f(t) = f_0 + t\,\Delta f_0 + \frac{t(t-1)}{2!}\,\Delta^2 f_0 + \frac{t(t-1)(t-2)}{3!}\,\Delta^3 f_0 + \cdots \qquad (2.1)$$

Gregory–Newton Backward-Difference Formula:

$$f(t) = f_0 + t\,\nabla f_0 + \frac{t(t+1)}{2!}\,\nabla^2 f_0 + \frac{t(t+1)(t+2)}{3!}\,\nabla^3 f_0 + \cdots \qquad (2.2)$$

where $f(t)$ = value of the function at time t
$\quad \Delta f_i = f_{i+1} - f_i$ [the first forward difference of $f(t)$ at period i]
$\quad \Delta^2 f_i = \Delta f_{i+1} - \Delta f_i$ [the second forward difference of $f(t)$ at period i]
$\quad \Delta^n f_i = \Delta^{n-1} f_{i+1} - \Delta^{n-1} f_i$ [the nth difference of $f(t)$ at period i]
$\quad \nabla f_i$ = the first backward difference of $f(t)$ at i

$$\nabla f_i = f_i - f_{i-1}$$
$$\nabla^n f_i = \nabla^{n-1} f_i - \nabla^{n-1} f_{i-1}$$

where f_0 is the value of the function at the baseline (time zero). The differences ∇f_i and Δf_i can easily be obtained from the difference table, as shown in Example 2-1.

EXAMPLE 2-1

Assume that the demand (in thousands of units) for a product x over the last 5 years is as given in Table 2-1. Determine the demand for next year.

Table 2-1. YEARLY DEMAND FOR THE PRODUCT

Year, t	0	1	2	3	4
Demand, $f(t)$	100	115	116	125	135

SOLUTION

We utilize Eqs. (2.1) and (2.2) in order to determine the demand for next year. Tables 2-2 and 2-3 are the forward- and backward-difference tables, respectively.

Table 2-2. FORWARD-DIFFERENCE TABLE

t	$f(t)$	Δf	$\Delta^2 f$	$\Delta^3 f$	$\Delta^4 f$
0	100	15	-14	22	-29
1	115	1	8	-7	
2	116	9	1		
3	125	10			
4	135				

From Eq. (2.1),

$$f(t) = 100 + 15t - \tfrac{14}{2}(t)(t-1) + \tfrac{22}{6}(t)(t-1)(t-2) - \tfrac{29}{24}(t)(t-1)(t-2)(t-3) \qquad (2.3)$$

To evaluate next year's demand, set $t = 5$:

$$f(5) = 110,000 \text{ units}$$

We can also use Eq. (2.2) to estimate next year's demand.

$$f(t) = f_0 + t\,\nabla f_0 + \frac{t(t+1)}{2!}\,\nabla^2 f_0 + \frac{t(t+1)(t+2)}{3!}\,\nabla^3 f_0 + \frac{t(t+1)(t+2)(t+3)}{4!}\,\nabla^4 f_0$$

$$f(t) = 135 + 10t + \tfrac{1}{2}t(t+1) - \tfrac{7}{6}t(t+1)(t+2) - \tfrac{29}{24}(t)(t+1)(t+2)(t+3) \qquad (2.4)$$

Set $t = 1$ in Eq. (2.4) to evaluate next year's demand.

$$f(1) = 110,000 \text{ units}$$

which is identical to the forecast obtained from Eq. (2.3). If the data in Table 2-1 are plotted, they will show an increasing trend of demand as time increases. Therefore, we should expect the demand of next year to be higher than 130,000 units. Due to the small increase in demand from year 1 to year 2, the forecasted demand from Eqs. (2.3) and (2.4) is much lower than the expected demand. This occurs because the underlying data-generating process contains some *noise or randomness*. If the demand of year 2

Table 2-3. BACKWARD-DIFFERENCE TABLE

t	$f(t)$	∇f	$\nabla^2 f$	$\nabla^3 f$	$\nabla^4 f$
0	100				
1	115	15			
2	116	1	-14		
3	125	9	8	22	
4	135	10	1	-7	-29

Table 2-4. FORWARD-DIFFERENCE TABLE

t	$f(t)$	Δf	$\Delta^2 f$	$\Delta^3 f$	Δ^4
0	100	15	-10	10	-5
1	115	5	0	5	
2	120	5	5		
3	125	10			
4	135				

were more nearly consistent with a smooth polynomial fit, say 120 instead of 116, we would obtain Table 2-4.

$$f(t) = 100 + 15t - 5(t)(t-1) + \tfrac{10}{6}(t)(t-1)(t-2) - \tfrac{5}{24}(t)(t-1)(t-2)(t-3) \quad (2.5)$$

and $f(5) = 150,000$ units, which is in closer agreement with what one would expect.

Equations (2.1) and (2.2) describe two polynomials that pass through every data point. These equations are poor tools for forecasting when the underlying data-generation process contains randomness or noise. Economic data, such as demand, are characterized by random variation. Therefore, we must introduce forecasting methods that account for randomness.

2.3.2. REGRESSION METHODS

The primary purpose of the Gregory–Newton method is to force the polynomial to assume the exact value of the tabulated functions at each of the points where the function is provided. If we assume the existence of randomness in the observed data that we are evaluating, a smooth function, such as a line, a logarithmic function, or a second-order polynomial, may be a more reasonable approximation of the average trend in the data. Such an approximation is shown in Fig. 2-1 for the linear case. The problem one is faced with in fitting such a function to the data is to determine the best criterion to use in judging the goodness of the fit. The *method of least squares* in regression models is known to be an efficient and unbiased criterion for fitting such functions. The method of least squares defines the best fit as that which minimizes the sum of squared errors between the observed data and the function. We will consider two regression models: (1) *simple linear regression* and (2) *multiple linear regression*.

Simple linear regression. Consider a set of data that may include extreme data points (or noise) due to some factors, as shown in Fig. 2-1. We are interested in finding a function that reflects the pattern in the data and reduces the errors to a minimum. The data in Fig. 2-1 suggest that the underlying data-generating process is a simple linear model with noise, which can be

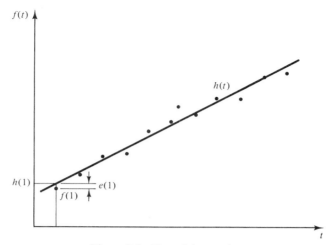

Figure 2-1. Plot of data points.

written as

$$f(t) = a_0 + a_1 t + \epsilon_t \tag{2.6}$$

where $f(t)$ = observed value of the function (dependent variable)
$\quad a_0, a_1$ = intercept and slope, respectively
$\quad t$ = time (independent variable)
$\quad \epsilon_t$ = random noise in the process at time t

It is assumed that ϵ_t is normally and independently distributed, with mean $\epsilon_t = 0$, var $(\epsilon_t) = \sigma^2$. Based on our assumption of a linear process with noise, we propose to fit a model of the form

$$h(t) = \hat{a}_0 + \hat{a}_1 t \tag{2.7}$$

where $\hat{a}_0$ and $\hat{a}_1$ are the estimates of a_0 and a_1, respectively. $h(t)$ = the value of the function forecasted at time t. Let $e(t) = h(t) - f(t)$ be the value of the error between the proposed polynomial $h(t)$ and the actual data $f(t)$. Then we define the sum of squares of errors, SS_E, as

$$SS_E = \sum_{t=1}^{n} e^2(t) \tag{2.8}$$

where n is the total number of data points used to estimate $h(t)$. We then obtain

$$SS_E = \sum_{t=1}^{n} [f(t) - h(t)]^2 \tag{2.9}$$

The minimization of SS_E is accomplished by taking the partial derivatives of SS_E with respect to the coefficients ($\hat{a}_0$ and $\hat{a}_1$) and setting the resulting equations equal to zero:

$$SS_E = \sum_{t=1}^{n} [f(t) - h(t)]^2 = \sum_{t=1}^{n} [f(t) - \hat{a}_0 - \hat{a}_1 t]^2$$

$$\frac{\partial SS_E}{\partial \hat{a}_0} = 2 \sum_{t=1}^{n} [f(t) - \hat{a}_0 - \hat{a}_1 t] = 0 \tag{2.10}$$

$$\frac{\partial SS_E}{\partial \hat{a}_1} = 2 \sum_{t=1}^{n} [f(t) - \hat{a}_0 - \hat{a}_1 t] t = 0 \tag{2.11}$$

Rewriting Eqs. (2.10) and (2.11) gives us

$$\sum_{t=1}^{n} f(t) = n\hat{a}_0 + \hat{a}_1 \sum_{t=1}^{n} t \tag{2.12}$$

$$\sum_{t=1}^{n} tf(t) = \hat{a}_0 \sum_{t=1}^{n} t + \hat{a}_1 \sum_{t=1}^{n} t^2 \tag{2.13}$$

which yields

$$\hat{a}_0 = \frac{\sum t^2 \sum f(t) - \sum t \sum tf(t)}{n \sum t^2 - (\sum t)^2} \tag{2.14}$$

$$\hat{a}_1 = \frac{n \sum tf(t) - \sum t \sum f(t)}{n \sum t^2 - (\sum t)^2} \tag{2.15}$$

We now use the simple linear regression model and the data in Example 2-1 to determine the forecast for next year's demand. Table 2-5 illustrates the necessary calculations.

$$\hat{a}_0 = \frac{30(591) - 10(1262)}{5(30) - (10)^2} = 102.2$$

$$\hat{a}_1 = \frac{5(1262) - 10(591)}{5(30) - (10)^2} = 8$$

The equation for $h(t)$ is then

$$h(t) = 102.2 + 8t$$

and the demand for next year is 142,200 units.

It is unlikely that the actual demand for the next year will be equal to the forecast, since the forecast is a point estimate based on an average expectation.

Table 2-5. DATA FOR A
LINEAR MODEL

t	$f(t)$	t^2	$tf(t)$
0	100	0	0
1	115	1	115
2	116	4	232
3	125	9	375
4	135	16	540
$\overline{10}$	$\overline{591}$	$\overline{30}$	$\overline{1262}$

It is reasonable to want to know something about the potential for error in the forecast. Since

$$h(t) = \hat{a}_0 + \hat{a}_1 t$$

and

$$f(t) = a_0 + a_1 t + \epsilon_t$$

the forecast error for time period $(t + Z)$ is

$$\begin{aligned} e(t + Z) &= h(t + Z) - f(t + Z) \\ &= (\hat{a}_0 - a_0) + (\hat{a}_1 - a_1)(t + Z) - \epsilon_{t+z} \end{aligned} \tag{2.16}$$

Equation (2.16) indicates that there are two sources of error in the forecast: the error due to the basic variance in $f(t)$ and the error due to incorrect estimation of the true intercept and slope. It can be shown that the standard error of the estimate for a_0 and a_1 are normally distributed, as is $e(t + Z)$. Since $e(t + Z)$ is a linear combination of normally distributed random variables, it too must be normally distributed. Hence the variance of the forecast can be written

$$\begin{aligned} \sigma_f^2 &= E[e(t + Z)^2] = E[(\hat{a}_0 - a_0)^2] + E[(\hat{a}_1 - a_1)^2](t + Z)^2 \\ &\quad + E[(\epsilon_{t+z})^2] + E[(\hat{a}_0 - a_0)(\hat{a}_1 - a_1)]2(t + Z) \end{aligned} \tag{2.17}$$

where E is the expected value operator. Equation (2.17) reduces to

$$\sigma_f^2 = \text{var}(\hat{a}_0) + 2(t + Z)\,\text{cov}(\hat{a}_0, \hat{a}_1) + (t + Z)^2\,\text{var}(\hat{a}_1) + \sigma^2 \tag{2.18}$$

It can be shown that

$$\text{var}(\hat{a}_0) = \sigma^2\,\frac{\sum t^2}{n \sum (t - \bar{t})^2} \tag{2.19}$$

$$\text{var} (\hat{a}_1) = \frac{\sigma^2}{\sum (t - \bar{t})^2} \tag{2.20}$$

$$\text{cov} (\hat{a}_0, \hat{a}_1) = \frac{-\bar{t}\sigma^2}{\sum (t - \bar{t})^2} \tag{2.21}$$

where $\bar{t} = \dfrac{1}{n} \sum t = $ the average value of t.

Substituting Eqs. (2.19), (2.20), and (2.21) into Eq. (2.18) and reducing yields

$$\sigma_f^2 = \sigma^2 \left\{ 1 + \frac{1}{n} + \frac{[(t + Z) - \bar{t}]^2}{\sum (t - \bar{t})^2} \right\} \tag{2.22}$$

Equation (2.22) could be used to compute the variance around the point estimate of the forecast if σ were known. However, the usual practice is to estimate the variance of the error term from the data, which requires the unbiased estimator

$$S^2 = \frac{1}{n - 2} \sum [f(t) - h(t)]^2 \tag{2.23}$$

Hence the operational form of estimating the forecast error variance is

$$S_f^2 = S^2 \left\{ 1 + \frac{1}{n} + \frac{[(t + Z) - \bar{t}]^2}{\sum (t - \bar{t})^2} \right\} \tag{2.24}$$

We note that the variance of the forecast error is reduced with a greater number of observations. Since S^2 follows the t-distribution, S_f^2 follows the t-distribution. The 95% confidence interval around the point estimate of the forecast is given by

$$h(t + Z) - t_{0.025} S_f \le f(t) \le h(t + Z) + t_{0.025} S_f$$

EXAMPLE 2-2

Using the data of Table 2-5, compute the 95% confidence interval around the forecast for the next time period.

SOLUTION

The calculations required for estimating S^2 and S_f^2 are shown in Table 2-6.

$$S^2 = \frac{1}{5 - 2} (34,800,000) = 11,600,000$$

$$S_f^2 = 11,600,000 \left[1 + \frac{1}{5} + \frac{(5 - 2)^2}{10} \right] = 24,360,000$$

Table 2-6. DATA FOR FORECAST VARIANCE

t	$(t - \bar{t})^2$	$f(t)$	$h(t)$	$[f(t) - h(t)]^2$
0	4	100,000	102,200	4,840,000
1	1	115,000	110,200	23,040,000
2	0	116,000	118,200	4,840,000
3	1	125,000	126,200	1,440,000
4	4	135,000	134,200	640,000
10	10	591,000	591,000	34,800,000

The standard error of the forecast estimate is

$$S_f = \sqrt{24,360,000} = 4936$$

Since $t_{0.025,\, 3\, df} = 3.182$, the 95% confidence interval around the point estimate of demand is

$$126,494 \le f(5) \le 157,906$$

Multiple linear regression. Although we illustrated the method of simple linear regression using a time-dependent model, the procedure is perfectly general for any independent variable. For example, if sales were deemed to be dependent on advertising, an appropriate model would replace time with advertising expenditure as the independent variable.

There are many practical situations where the dependent variable is a function of more than one independent variable. For example, the sales of a product may depend on the average income of the customer in a given area (x_1), the amount of dollars spent on advertising (x_2), and so on. In this case, the dependent variable (sales) can be expressed as a function of more than one independent variable. Consider the following model with two independent variables:

$$h(x_{1i}, x_{2i}) = \hat{a}_0 + \hat{a}_1 x_{1i} + \hat{a}_2 x_{2i} \tag{2.25}$$

which estimates the process

$$f(x_{1i}, x_{2i}) = a_0 + a_1 x_{1i} + a_2 x_{2i} + \epsilon_i \tag{2.26}$$

where i indexes the ith observation. Following the development in Eqs. (2.9), (2.10), and (2.11), we obtain three linear equations in $\hat{a}_0$, $\hat{a}_1$, and $\hat{a}_2$. These equations are

$$\sum_{i=1}^{n} f(x_{1i}, x_{2i}) = n\hat{a}_0 + \hat{a}_1 \sum_{i=1}^{n} x_{1i} + \hat{a}_2 \sum_{i=1}^{n} x_{2i} \tag{2.27}$$

$$\sum_{i=1}^{n} x_{1i} f(x_{1i}, x_{2i}) = \hat{a}_0 \sum_{i=1}^{n} x_{1i} + \hat{a}_1 \sum_{i=1}^{n} x_{1i}^2 + \hat{a}_2 \sum_{i=1}^{n} x_{1i} x_{2i} \qquad (2.28)$$

$$\sum_{i=1}^{n} x_{2i} f(x_{1i}, x_{2i}) = \hat{a}_0 \sum_{i=1}^{n} x_{2i} + \hat{a}_1 \sum_{i=1}^{n} x_{1i} x_{2i} + \hat{a}_2 \sum_{i=1}^{n} x_{2i}^2 \qquad (2.29)$$

Equations (2.27), (2.28), and (2.29) can be solved simultaneously to obtain the coefficients $\hat{a}_0$, $\hat{a}_1$, and $\hat{a}_2$.

EXAMPLE 2-3

For the data given in Example 2-1, determine the forecast of demand for next year assuming a quadratic model of the form

$$h(t, t^2) = \hat{a}_0 + \hat{a}_1 t + \hat{a}_2 t^2$$

SOLUTION

The quadratic model is a special form of a multiple regression model. Let $x_1 = t$ and $x_2 = t^2$. Table 2-7 gives the necessary calculations.

$$591 = 5\hat{a}_0 + 10\hat{a}_1 + 30\hat{a}_2$$

$$1262 = 10\hat{a}_0 + 30\hat{a}_1 + 100\hat{a}_2$$

$$3864 = 30\hat{a}_0 + 100\hat{a}_1 + 354\hat{a}_2$$

$$\hat{a}_0 = 101.91 \qquad \hat{a}_1 = 8.571 \qquad \hat{a}_2 = -0.1428$$

In terms of the original variables, the quadratic forecasting model is

$$h(t, t^2) = 101.91 + 8.571t - 0.1428t^2 \qquad (2.30)$$

Set $t = 5$ in Eq. (2.30) to get next year's demand of 141,195 units.

As in the case of simple linear regression, it is not sufficient to know only the point estimate of demand; one should also know the potential for error around that point estimate. Leaving the derivation to advanced texts, we offer

Table 2-7. CALCULATIONS FOR QUADRATIC MODEL

$f(x_{1i}, x_{2i})$	x_{1i}	x_{2i}	$x_{1i} f(x_{1i}, x_{2i})$	$x_{2i} f(x_{1i}, x_{2i})$	x_{1i}^2	x_{2i}^2	x_{1i}, x_{2i}
100	0	0	0	0	0	0	0
115	1	1	115	115	1	1	1
116	2	4	232	464	4	16	8
125	3	9	375	1125	9	81	27
135	4	16	540	2160	16	256	64
591	10	30	1262	3864	30	354	100

without proof the following equation for the variance of the forecast error in multiple linear regression:

$$S_f^2 = S^2[1 + \tilde{X}(X'X)^{-1}\tilde{X}']$$

where $\tilde{X}$ = vector of forecasted observations on the m independent variables, $\tilde{X}'$ is its transposition

$X = n \times m$ matrix of observations on the independent variables, X' is its transposition

EXAMPLE 2-4

For the data used in Example 2-3, determine the 95% confidence interval around the forecast of demand for year 5. Calculations are shown in Table 2-8.

SOLUTION

$$S^2 = \frac{1}{n-3} \sum [f(x_1 x_2) - h(x_1 x_2)]^2 = \tfrac{1}{2}(34.56) = 17.28$$

$$X'X = \begin{bmatrix} 0 & 1 & 2 & 3 & 4 \\ 0 & 1 & 4 & 9 & 16 \end{bmatrix} \begin{bmatrix} 0 & 0 \\ 1 & 1 \\ 2 & 4 \\ 3 & 9 \\ 4 & 16 \end{bmatrix} = \begin{bmatrix} 30 & 100 \\ 100 & 354 \end{bmatrix}$$

$$[X'X]^{-1} = \begin{bmatrix} 0.57097 & -0.16129 \\ -0.16129 & 0.04839 \end{bmatrix}$$

For year 5,

$$\tilde{X} = [5 \quad 25]$$

$$S_f^2 = 17.28\left(1 + [5 \quad 25]\begin{bmatrix} 0.57097 & -0.16129 \\ -0.16129 & 0.04839 \end{bmatrix}\begin{bmatrix} 5 \\ 25 \end{bmatrix}\right)$$

$$S_f^2 = 89.74$$

Table 2-8. CALCULATIONS FOR EXAMPLE 2-4

x_{1i}	x_{2i}	$f(x_{1i} x_{2i})$	$h(x_{1i} x_{2i})$	$[f(x_{1i} x_{2i}) - h(x_{1i} x_{2i})]^2$
0	0	100	101.91	3.65
1	1	115	110.34	21.72
2	4	116	118.48	6.20
3	9	125	126.34	1.80
4	16	135	133.91	1.19
10	30	591	590.98	34.56

Therefore, $S_f = 9.473$ and the 95% confidence interval is given by

$$f(x_1 x_2) \pm t_{0.025,2} S_f$$

$$141,195 - (4.303)(9.473) \leq f(t, t^2) \leq 141,195 + (4.303)(9.473)$$

$$100,432 \leq f(t, t^2) \leq 181,957$$

Linear regression and nonlinear functions. In many situations there are nonlinear relationships between the dependent and independent variables which can be expressed in a linear model. To use the linear regression model, it becomes necessary to transform the function to fit the linear regression model. Some of these nonlinear functions and their transformations are given below.

$$h(t) = AB^t \tag{2.31}$$

Taking the logarithm of both sides of Eq. (2.31) yields

$$\log h(t) = \log A + (t) \log B \tag{2.32}$$

Substituting $H(t) = \log h(t)$, $\hat{a}_0 = \log A$, and $\hat{a}_1 = \log B$ in Eq. (2.32), then

$$H(t) = \hat{a}_0 + \hat{a}_1 t \tag{2.33}$$

Equations (2.7) and (2.33) are identical. Values of $\hat{a}_0$ and $\hat{a}_1$ can be found as presented earlier. Similarly, if the function is represented by

$$h(t) = Ae^{Bt}$$

it can be transformed into the linear form $\log h(t) = \log A + Bt$. Substitution of $H(t) = \log h(t)$, $\hat{a}_0 = \log A$, and $\hat{a}_1 = B$ yields an equation identical to Eq. (2.33). A function of the form

$$h(t) = \frac{t}{At - B}$$

can be linearized by substituting

$$H(t) = \frac{1}{h(t)} \quad \text{and} \quad t' = \frac{1}{t}$$

to get

$$H(t) = A - Bt'$$

The demand, sales, and production rate of a product may display periodic phenomena, which add another dimension to the forecasting problem. When these phenomena exist, trigonometric models of the *Fourier series* type can be used to forecast the demand and sales at any given time (t). Snow tires, antifreeze, and air conditioners are examples of products that show a periodic change in the demand and consequently the production rate.

EXAMPLE 2-5

As an example of the forecasting procedures for seasonal variations (periodic changes), the number of snow tires sold during 1 year at a tire distributor is shown in Table 2-9. The data are plotted in Fig. 2-2.

Table 2-9. SNOW TIRES SOLD

Month t	Number of Tires, $f(t)$
Jan.	200
Feb.	250
Mar.	300
Apr.	280
May	240
June	210
July	180
Aug.	150
Sept.	120
Oct.	110
Nov.	170
Dec.	195

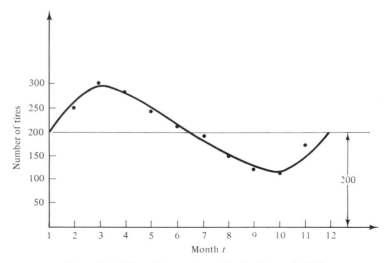

Figure 2-2. Plot of snow tire sales for Example 2-5.

The shape of the data suggests the use of a trigonometric function. We might, for example, assume a function of the form

$$h(t) = A + B \sin \pi t/6 \qquad (2.34)$$

where A = origin of the sine wave
B = amplitude
$\pi t/6$ = periodicity (frequency), radians

In general, frequency is specified by $\sin (2\pi t/n)$, where n is the number of observations in a seasonal pattern.

We have specified the period of the sine as 12; that is, the seasonal pattern is repeated every 12 observations. This would appear to be reasonable from an examination of the data. Applying the same least-squares method as described earlier for polynomials, we define

$$SS_E = \sum_{t=1}^{n} \left[\hat{A} + \hat{B} \sin \frac{\pi t}{6} - f(t) \right]^2 \qquad (2.35)$$

where $\hat{A}$ and $\hat{B}$ are the parameters that minimize SS_E. Taking the partial derivative of Eq. (2.35), with respect to $\hat{A}$ and $\hat{B}$ results in Eqs. (2.36) and (2.37), respectively.

$$\frac{\partial E}{\partial \hat{A}} = 2 \sum_{t=1}^{n} \left[\hat{A} + \hat{B} \sin \frac{\pi t}{6} - f(t) \right] = 0 \qquad (2.36)$$

$$\frac{\partial E}{\partial \hat{B}} = 2 \sum_{t=1}^{n} \left[\hat{A} + \hat{B} \sin \frac{\pi t}{6} - f(t) \right] \sin \frac{\pi t}{6} = 0 \qquad (2.37)$$

The equations above can be written as

$$\sum_{t=1}^{n} f(t) = n\hat{A} + \hat{B} \sum_{t=1}^{n} \sin \frac{\pi t}{6} \qquad (2.38)$$

$$\sum_{t=1}^{n} f(t) \sin \frac{\pi t}{6} = \hat{A} \sum_{t=1}^{n} \sin \frac{\pi t}{6} + \hat{B} \sum_{t=1}^{n} \sin^2 \frac{\pi t}{6} \qquad (2.39)$$

The parameters $\hat{A}$ and $\hat{B}$ can be obtained by simultaneously solving Eqs. (2.38) and (2.39). For this particular example ($n = 12$), Table 2-10 is constructed. Substituting in Eqs. (2.38) and (2.39) yields

$$h(t) = 200.41 + 76.45 \sin \frac{\pi t}{6}$$

Table 2-10. CALCULATIONS FOR EXAMPLE 2-5

t	$f(t)$	$\sin \dfrac{\pi t}{6}$	$f(t) \sin \dfrac{\pi t}{6}$	$\sin^2 \dfrac{\pi t}{6}$
1	200	0.5000	100	0.2500
2	250	0.8660	216.5	0.7499
3	300	1.000	300	1.00
4	280	0.8660	242.4	0.7499
5	240	0.500	120.0	0.25
6	210	0.000	0.0	0.0
7	180	−0.500	−90.0	0.25
8	150	−0.8660	−129.9	0.7499
9	120	−1.000	−120.0	1.00
10	110	−0.8660	−95.26	0.7499
11	170	−0.500	−85.0	0.25
12	195	0.000	0.0	0.0
78	2405	0	485.74	6

Correlation. Regression is used to fit a function through a set of data by the criterion of least squares. However, regression tells us nothing about how well the function fits the data. For this we need the correlation coefficient.

Simple correlation, r, is a measure of how well two variables move together. The numerical value of r lies between $+1$ and -1; it is positive if $f(x)$ increases as x increases and negative if $f(x)$ decreases as x increases. If $r = 0$, no relationship exists between $f(x)$ and x. Figure 2-3 shows these cases.

It can be shown that the least-squares line always passes through the average values of $f(x)$ and x; that is $(\overline{f(x)}, \bar{x})$. This occurs even if this point is not itself a data point. Given x, the expected value of $f(x)$ [i.e., $h(x)$] is given by the function

$$h(x) = \overline{f(x)} + \hat{a}_1(x - \bar{x})$$

In the absence of knowing the relationship defined by $h(x)$, if one were given a value for x, the *best-guess* estimate one could make of $f(x)$ would be $\overline{f(x)}$, the mean. Hence, of the total variation of $f(x)$ in the data set, $\hat{a}_1(x - \bar{x})$ is the amount accounted for by the functional relationship. This is illustrated by Fig. 2-4.

Recall that in applying the regression method, we minimized the residual sum of squares to fit the function to the data. We redefine the residual sum of squares.

$$
\begin{aligned}
SS_E &= \sum_i [f(x_i) - \hat{a}_0 - \hat{a}_1 x_i]^2 \\
&= \sum_i [f(x_i) - \overline{f(x_i)} - \hat{a}_1(x_i - \bar{x})]^2
\end{aligned}
$$

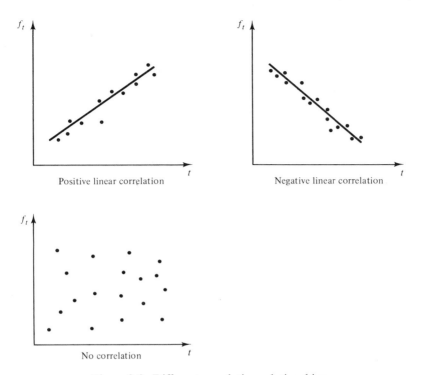

Figure 2-3. Different correlation relationships.

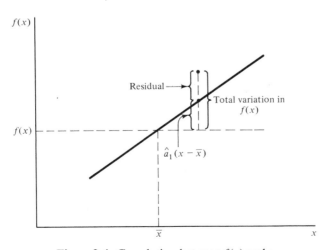

Figure 2-4. Correlation between $f(x)$ and x.

Expanding the right-hand side, we obtain

$$SS_E = \sum [f(x_i) - \overline{f(x_i)}]^2 + \hat{a}_1^2 \sum (x_i - \bar{x})^2 - 2\hat{a}_1 \sum [f(x_i) - \overline{f(x_i)}](x_i - \bar{x})$$

$$(2.40)$$

We define the following sum of squares:

$$S_{ff} = \sum [f(x_i) - \overline{f(x_i)}]^2$$
$$S_{xx} = \sum (x_i - \bar{x})^2$$
$$S_{xf} = \sum [f(x_i) - \overline{f(x_i)}](x_i - \bar{x}) = \sum f(x_i)(x_i - \bar{x})$$

Substituting into Eq. (2.40) yields

$$SS_E = S_{ff} + \hat{a}_1^2 S_{xx} - 2\hat{a}_1 S_{xf}$$

From Eq. (2.15) we know that

$$\hat{a}_1 = \frac{n \sum_i x_i f(x_i) - \sum_i x_i \sum_i f(x_i)}{n \sum_i x_i^2 - \left(\sum_i x_i\right)^2} = \frac{\sum_i f(x_i)(x_i - \bar{x})}{\sum_i (x_i - \bar{x})^2}$$

which when substituted into Eq. (2.40) yields

$$SS_E = S_{ff} + \left(\frac{S_{xf}}{S_{xx}}\right)^2 S_{xx} - 2\left(\frac{S_{xf}}{S_{xx}}\right)S_{xf}$$

$$SS_E = S_{ff} - \frac{S_{xf}^2}{S_{xx}}$$

$$SS_E = S_{ff} - \hat{a}_1 S_{xf}$$

where $\quad S_{ff}$ = measure of the total variation in $f(x)$
$\hat{a}_1 S_{xf}$ = measure of the amount of the total variation explained by the regression line
$SS_E = S_{ff} - \hat{a}_1 S_{xf}$ = measure of the residual, or unexplained variation
r^2 = coefficient of determination, defined as the amount of total variation in the data that is explained by the regression line

For the simple linear regression

$$r^2 = 1 - \frac{SS_\epsilon}{S_{ff}} = 1 - \frac{S_{ff} - \hat{a}_1 S_{xf}}{S_{ff}} = \frac{\hat{a}_1 S_{xf}}{S_{ff}}$$

For the case of multiple regression, when r^2 is generalized to the k-independent-variable case,

$$r^2 = \frac{\hat{a}_1 S_{x_1 f} + \hat{a}_2 S_{x_2 f} + \cdots + \hat{a}_k S_{x_k f}}{S_{ff}}$$

The square root of the coefficient of determination is the correlation coefficient.

EXAMPLE 2-6

Using the data of Examples 2-1 and 2-2, compare the coefficient of determination with the coefficient of multiple determination. Calculations are shown in Table 2-11.

Table 2-11. CALCULATIONS FOR EXAMPLE 2-6

x_{1i}	x_{2i}	$f(\cdot)^*$	$(x_{1i} - \bar{x}_1)^2$	$(x_{2i} - \bar{x}_2)^2$	$(x_{1i} - \bar{x}_1)f(\cdot)$	$(x_{2i} - \bar{x}_2)f(\cdot)$	$[f(\cdot) - \bar{f}(\cdot)]^2$
0	0	100	4	36	-200	-600	331.24
1	1	115	1	25	-115	-575	10.24
2	4	116	0	4	0	-232	4.84
3	9	125	1	9	125	375	46.24
4	16	135	4	100	270	1350	282.24
$\overline{10}$	$\overline{30}$	$\overline{591}$	$\overline{10}$	$\overline{174}$	$\overline{80}$	318	$\overline{674.8}$

*$f(\cdot)$ denotes $f(x_{1i}, x_{2i})$.

SOLUTION

$$r^2_{f x_1} = \frac{8(80)}{674.8} = 0.948$$

$$r^2_{f x_1 x_2} = \frac{(8.571)(80) + (-0.1428)(318)}{674.8} = 0.949$$

There is an important observation that we can make concerning the results of Example 2-6. Note that x_1 explains a great amount of the variability in $f(x)$; the addition of x_2 adds virtually nothing. It is always possible to raise the value of r^2 in a regression by bringing new explanatory variables into the model. However, those variables may not add much to the reduction in SS_E, and in fact, may widen the confidence interval around the forecast. Recall that the distribution of the forecast error is a function of the noise in the data-generating process and the error in the specification of the population parameters [more specifically, the standard error in the parameters (a_i)]. Hence, adding explanatory variables with large standard errors may increase the forecast error distribution. This is what occurs in this case, when our model $h(x_1)$ goes to $h(x_1 x_2)$ and accounts for the increase in the range of the 95% confidence interval.

2.3.3. MOVING-AVERAGE METHODS

The regression method assumes that a relationship exists between the independent variables and the dependent variable and that this relationship is stable over time. In order to use regression, the values of the independent variables in the forecast must be known. If, as in some of our examples, the independent variable is time, this does not create an especially difficult problem. However, it would be naive to expect economic variables, such as sales, to be a simple function of time. It is more likely that sales would be a function of some other economic variable, such as gross national product (GNP) or disposable personnel income (DPI). This fact often leads to the difficulty of trying to forecast a sales level using projections of GNP or PDI, which are themselves forecasts.

A class of much easier to use *tracking* models that do not require an exogeneous independent variable is the moving-average model. The moving-average model tracks the changing movement of the variable of interest as a function of its prior levels.

Simple moving averages. When the data are stable, we may use the *simple average* as a forecasting method. The simple average is defined as follows:

$$\text{simple average} = \frac{\sum_{t=1}^{n} f(t)}{n} \qquad (2.41)$$

where n is the total number of data points available.

If there is a trend in the data, one may use the simple average as an initial value of the forecast, which is then adjusted according to the trend. The effect of trend may be included in the average directly by using the *moving average*. A moving average is obtained by summing and averaging the values from a given number of data points repetitively. Each time the moving average is calculated, the oldest data point is deleted and a new data point is added. In other words, we select a period over which the data are averaged. Of course, the length of the period is less than the total number of available periods. Also, the length of the period should not be long; otherwise, the effect of a temporary increase (or decrease) in a data point will be diminished. However, moving averages calculated over short periods will be *very sensitive* to changes in the data. The moving average is calculated as

$$\text{moving average} = \frac{\sum_{t=1}^{n} f(t)}{n} \qquad (2.42)$$

where n is the number of data points that define the length of the period.

EXAMPLE 2-7

Use the moving average (MA) method to determine next year's demand for the data given below.

Year	0	1	2	3	4	5
Demand	106	110	107	105	115	112

SOLUTION

Let us assume that 3-year and 5-year moving averages are used. Note that the moving average is recorded (Table 2-12) in the center position of the data it averages. For example, the 3-year moving average for $t = 0, 1, 2$ is $(106 + 110 + 107)/3$ and is recorded opposite to $t = 1$.

Based on the 3-year moving average, the forecasted demand for next year is 110.60. The five-year moving average yields 109.2 as the demand for next year. In this example, the 5-year moving average does not show the increase in the demand for the last 2 years, however, the 3-year moving average shows the effect of this increase in the demand.

Table 2-12. MOVING AVERAGES

t	$f(t)$	3-year MA	5-year MA
0	106		
1	110	107.6	
2	107	107.3	108.6
3	105	109.0	109.2
4	115	110.6	
5	112		

Sometimes, it is desirable to place more emphasis on the recent data. This can be achieved by assigning more weight, w_t, to these data, and the weighted moving average becomes

$$\text{weighted moving average} = \frac{\sum W_t f_t}{\sum W_t} \tag{2.43}$$

Moving average for data with seasonal demands. In this section we illustrate the use of the moving average in forecasting for data with seasonal demand (we ignore the effect of general trends for the moment). This can best be explained by the following example.

EXAMPLE 2-8

Apply the method of moving averages to the seasonal data on the number of snow tires sold for the first 4 years, as shown in Table 2-13. Figure 2-5 shows that sales reach highest values in March, while its lowest values are in September. Since the annual sales are constant (no significant change in the total sales), we can assume that there is no general trend for the data.

Table 2-13. SNOW TIRE SALES OVER 4 YEARS

| Month | Number of Tires Sold | | | |
	Year 1	Year 2	Year 3	Year 4
Jan.	200	190	195	205
Feb.	250	245	250	245
Mar.	300	290	295	305
Apr.	280	275	270	275
May	240	245	250	245
June	210	205	200	205
July	180	175	180	175
Aug.	150	145	140	155
Sept.	120	120	125	115
Oct.	110	115	110	115
Nov.	170	175	180	165
Dec.	195	200	195	190
	2405	2380	2390	2395

In order to show the seasonal cycle, one may plot 3-month, 6-month, and 12-month moving totals for sales. The N-month moving total is calculated as

$$N\text{-month moving total for month } i = \sum_{k=1}^{N} S_{i-k+1}$$

where S_j is the sales during month j. For example, the 6-month moving total for the months of

$$\text{July (year 1)} = 180 + 210 + 240 + 280 + 300 + 250 = 1460$$

$$\text{January (year 2)} = 190 + 195 + 170 + 110 + 120 + 150 = 935$$

The 3-month moving totals are given in Table 2-14 and those for 6-month and 12-month moving totals are given in Table 2-15.

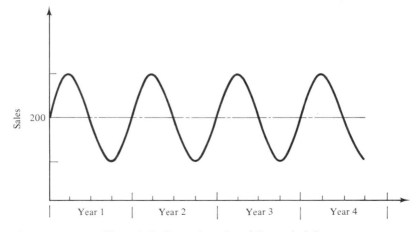

Figure 2-5. Snow tire sales of Example 2-8.

Table 2-14. THREE-MONTH MOVING TOTALS

	3-Month Moving Totals				
Month	*Year 1*	*Year 2*	*Year 3*	*Year 4*	*Average*
Jan.	—	555	570	580	568
Feb.	—	630	645	645	640
Mar.	750	725	740	755	743
Apr.	830	810	815	825	820
May	820	810	815	825	818
June	730	725	720	725	725
July	630	625	630	625	628
Aug.	540	525	520	535	530
Sept.	450	440	445	445	445
Oct.	380	380	375	385	380
Nov.	400	410	415	395	405
Dec.	475	490	485	470	480

Upon plotting the data from Tables 2-14 and 2-15, as shown in Figs. 2-6, 2-7, and 2-8, respectively, it is concluded that the moving total removes any monthly changes and shows the seasonality of the data. This seasonality is more evident in Fig. 2-6 due to the small number of periods used in calculating the moving total ($N = 3$); however, it is less evident in Fig. 2-7 and completely diminishes in Fig. 2-8. This implies that the choice of the number of periods for calculating the moving totals is of extreme importance in detecting the seasonality. Moving totals based on a larger number of periods can be used to show the general trend in the data.

In practice, a 3-month moving total is usually used to generate the seasonal cycles in the data. Two moving total approaches can be used for forecasting.

—*Method 1:* The forecast is based on the average of each month of the past n years (the simple average) where n is the number of years for which data are available.

—*Method 2:* The forecast is based on the average N-month moving total.

Table 2-15. SIX-MONTH AND TWELVE-MONTH MOVING TOTALS

	6-Month Moving Totals					12-Month Moving Totals				
Month	*Yr 1*	*Yr 2*	*Yr 3*	*Yr 4*	*Avg.*	*Yr 1*	*Yr 2*	*Yr 3*	*Yr 4*	*Avg.*
Jan.	—	935	950	955	947	—	2395	2385	2395	2388
Feb.	—	1030	1055	1060	1048	—	2390	2390	2395	2392
Mar.	—	1200	1230	1240	1190	—	2380	2395	2405	2395
Apr.	—	1365	1385	1405	1385	—	2375	2390	2400	2388
May	—	1440	1460	1470	1457	—	2380	2395	2405	2393
June	1480	1450	1460	1480	1468	—	2375	2390	2410	2391
July	1460	1435	1445	1450	1448	—	2370	2395	2405	2390
Aug.	1360	1335	1335	1360	1348	—	2365	2390	2420	2391
Sept.	1180	1165	1165	1170	1170	—	2365	2395	2410	2390
Oct.	1010	1005	1005	1010	1008	—	2370	2390	2415	2391
Nov.	940	935	935	930	935	—	2385	2395	2400	2393
Dec.	925	930	930	915	925	2405	2380	2390	2395	2393

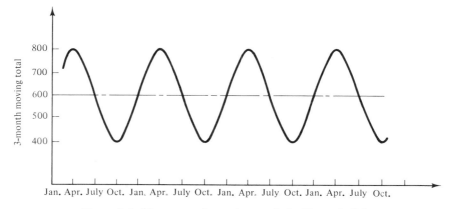

Figure 2-6. Three-month moving totals for Example 2-8.

The choice between method 1 and method 2 depends on the change in the monthly data. For example, if the monthly demands are independent, method 1 should be used. However, method 2 is adopted if there is dependency between demands of previous months and those of next months. These methods (1 and 2) are used for forecasting, as shown in Tables 2-16 and 2-17, for 3-month and 6-month moving totals, respectively. From Tables 2-16 and 2-17 it can be concluded that the differences between methods 1 and 2 are insignificant for this set of data.

Analysis of trend in seasonal demand. In order to include the effect of trend in forecasting for data with seasonal demand, it is recommended to plot the data to show the existence of trend (linear or nonlinear, increasing or

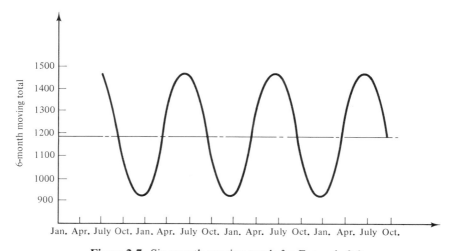

Figure 2-7. Six-month moving totals for Example 2-8.

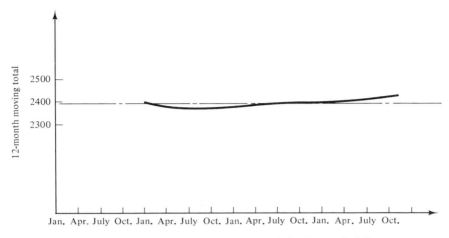

Figure 2-8. Twelve-month moving totals for Example 2-8.

decreasing). The data of a 12-month moving total are shown in Fig. 2-9; it is noted that there is an upward (increasing) trend with an increment of Δ units per year. This means that the demand of each month will exceed the corresponding month in the preceding year by Δ units. The demand record for the past 3 years is given in Table 2-18, where $d_{i,\,j}$ = demand in the ith month for the jth year. If the trend increases by the same rate (trend remains stationary), this record would read in terms of the next year's demand [8], as shown in Table 2-19.

Table 2-16. USING METHODS 1 AND 2 WITH 3-MONTH DEMAND

Month	Method 1: Monthly Forecast Based on Simple Average (A)	Method 2			
		3-Month Moving Total (B)	Previous 2-Month Demand (C)	Forecast (D) = (B) − (C)	Difference (A) − (D)
Jan.	198	568	367	201	−3
Feb.	248	640	393	247	+1
Mar.	298	743	446	297	+1
Apr.	275	820	546	274	+1
May	245	818	573	245	0
June	205	725	520	205	0
July	178	628	550	178	0
Aug.	148	530	383	147	+1
Sept.	120	445	326	119	+1
Oct.	113	380	268	112	+1
Nov.	172	405	233	172	0
Dec.	195	480	285	195	0

Table 2-17. METHODS 1 AND 2 USING 6-MONTH MOVING TOTAL

Month	Method 1: Monthly Forecast Based on Simple Average (A)	Method 2			
		6-Month Moving Total (B)	Previous 5-Month Demand (C)	Forecast (D) = (B) − (C)	Difference (A) − (D)
Jan.	198	947	748	199	−1
Feb.	248	1048	798	250	−2
Mar.	298	1223	926	297	+1
Apr.	275	1385	1111	274	+1
May	245	1457	1214	243	+2
June	205	1468	1264	204	+1
July	178	1448	1271	177	+1
Aug.	148	1348	1201	147	+1
Sept.	120	1170	1051	119	+1
Oct.	113	1008	896	112	+1
Nov.	172	935	764	171	+1
Dec.	195	925	731	194	+1

In fact, we have translated the demands to next year's level; hence the forecast based on the monthly average of previous years is given by D_1 (forecast for January of next year), and D_i is the forecast for the ith month of next year. When the existing record consists of n previous years, the forecast becomes

$$D_j = \frac{1}{n} \sum_{j=1}^{n} d_{i,j} + \frac{n+1}{2} \Delta \qquad (2.44)$$

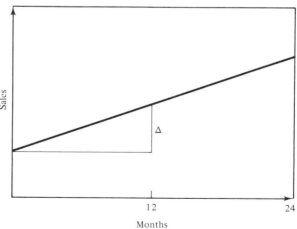

Figure 2-9. Increasing trend in the data of Δ units per year.

Table 2-18. DEMAND RECORD FOR PAST 3 YEARS

Month	First Year	Second Year	Third Year
Jan.	$d_{1,1}$	$d_{1,2}$	$d_{1,3}$
Feb.	$d_{2,1}$	$d_{2,2}$	$d_{2,3}$
Mar.	$d_{3,1}$	$d_{3,2}$	$d_{3,3}$
$\vdots$	$\vdots$	$\vdots$	$\vdots$
i	$d_{i,1}$	$d_{i,2}$	$d_{i,3}$
$\vdots$	$\vdots$	$\vdots$	$\vdots$
Dec.	$d_{12,1}$	$d_{12,2}$	$d_{12,3}$

Table 2-19. NEXT YEAR'S DEMAND

Month	First Year	Second Year	Third Year	Average
Jan.	$d_{1,1} + 3\Delta$	$d_{1,2} + 2\Delta$	$d_{1,3} + \Delta$	$D_1 = \frac{1}{3} \sum_{j=1}^{3} d_{1,j} + 2\Delta$
Feb.	$d_{2,1} + 3\Delta$	$d_{2,2} + 2\Delta$	$d_{2,3} + \Delta$	$D_2 = \frac{1}{3} \sum_{j=1}^{3} d_{2,j} + 2\Delta$
$\vdots$	$\vdots$	$\vdots$	$\vdots$	$\vdots$ $\quad$ $\vdots$
Dec.	$d_{12,1} + 3\Delta$	$d_{12,2} + 2\Delta$	$d_{12,3} + \Delta$	$D_{12} = \frac{1}{3} \sum_{j=1}^{3} d_{12,j} + 2\Delta$

Other types of trends (nonlinear, cyclic, etc.) can be dealt with in very much the same way.

2.3.4. EXPONENTIAL SMOOTHING

This is one of the most popular techniques of time-series analysis. As we have indicated earlier, time-series analysis identifies the historical pattern of the data and projects this into the future. In using the forecasting techniques, four main components should be identified: *trend, cyclical, seasonal,* and *irregular components. Trend* was presented earlier as the long-run historical component of the series (upward or downward). *Seasonal* variations have cycles that are repeated annually and these variations have a range of 1 year (see Examples 2-5 and 2-8). *Cycles* usually have a longer range than 1 year; they can be attributed to long-term business cycles in the economy. Unemployment, interest rates, and inflation rates are examples of dependent variables which exhibit cyclical variations. The *irregular components,* or unexplained variations, represent variations which are not caused by cyclical, trend, and seasonal components. Figure 2-10 shows the trend, cyclical, and seasonal components of forecasting models.

Exponential smoothing is a mathematical technique that utilizes the same principle of an N-period moving average; however, it requires fewer calculations than does the moving average. In addition, it does not require keeping

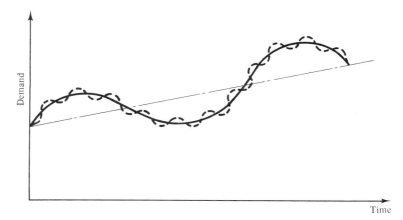

Figure 2-10. Trend, cyclical, and irregular components. Trend, dashed line; cycles, solid line; seasonal component, dotted line.

the historical data over a long period of time; it requires only recent data. Exponential smoothing is especially useful for short-term forecasting.

Simple exponential smoothing. This method is used for forecasting when the data are locally constant (i.e., the data have an insignificant trend). The smoothing characteristic is controlled by using a factor called the *smoothing factor* α; its value is between 0 and 1 inclusive. The purpose of this factor is to place more emphasis on recent data.

The forecasting of demand by utilizing exponential smoothing results from the equation

$$h_t = \alpha f_t + (1 - \alpha)h_{t-1} \tag{2.45}$$

where h_t = new estimate (forecast) of demand made in period t
f_t = actual value of the demand for period t
h_{t-1} = estimate of the demand made in period $t - 1$
α = smoothing constant $(0 \le \alpha \le 1)$

Equation (2.45) can be interpreted as follows: A smoothed forecast (h_t) equals a fraction α of the actual demand f_t *plus a fraction* $(1 - \alpha)$ of the estimate of the demand (h_{t-1}) during the previous period. The initial value of h_{t-1} can be estimated by using one of the following approaches:

1. When past data are available, h_{t-1} = summation of the most recent (n) data points divided by n.
2. The initial value of f_{t-1}.
3. When data are not available, we can use judgment.

We should be careful in selecting the value of α. Brown [2] has suggested a practical range of α to be $0.01 \le \alpha \le 0.3$. It can be proven that forecasting by using simple exponential smoothing will yield the same values as those of the simple moving average if $\alpha = 2/(n + 1)$, where n is the number of observations used in calculating the moving average. Let the initial forecast at $t = 0$ be h_0; then

$$h_1 = \alpha f_1 + (1 - \alpha)h_0 \qquad (2.46)$$

$$h_2 = \alpha f_2 + (1 - \alpha)h_1 \qquad (2.47)$$

Substitute the value of h_1 from Eq. (2.46) into Eq. (2.47) to get

$$h_2 = \alpha f_2 + \alpha(1 - \alpha)f_1 + (1 - \alpha)^2 h_0 \qquad (2.48)$$

$$h_3 = \alpha f_3 + (1 - \alpha)h_2 \qquad (2.49)$$

Substituting Eq. (2.48) into Eq. (2.49) yields

$$h_3 = \alpha f_3 + \alpha(1 - \alpha)f_2 + \alpha(1 - \alpha)^2 f_1 + (1 - \alpha)^3 h_0 \qquad (2.50)$$

In general,

$$h_t = \alpha f_t + (1 - \alpha)h_{t-1} \qquad (2.51)$$

or

$$h_t = \alpha f_t + \alpha(1 - \alpha)f_{t-1} + \alpha(1 - \alpha)^2 f_{t-2} + \cdots$$
$$+ \alpha(1 - \alpha)^{t-1} f_1 + (1 - \alpha)^t h_0 \qquad (2.52)$$

EXAMPLE 2-9

Use the simple exponential smoothing technique to forecast the demand for next year for the data given in Example 2-1.

SOLUTION

Assume that $\alpha = 0.3$ and using Eq. (2.52); then

$$h_4 = \alpha f_4 + \alpha(1 - \alpha)f_3 + \alpha(1 - \alpha)^2 f_2 + \alpha(1 - \alpha)^3 f_1 + (1 - \alpha)^4 h_0$$
$$= (0.3)(135) + (0.3)(0.7)(125) + (0.3)(0.7)^2(116) + (0.3)(0.7)^3(115)$$
$$+ (0.7)^4(118.2) = 124.015$$
$$= 124{,}015 \text{ units}$$

It should be noted that h_0 is the simple average of the data.

Let $\alpha = 0.8$; then

$$h_4 = (0.8)(135) + (0.8)(0.2)(125) + (0.8)(0.2)^2(116) + (0.8)(0.2)^3(115)$$
$$+ (0.2)^4(118.2) = 132.637$$
$$= 132{,}637 \text{ units}$$

As shown above the forecasted demand is sensitive to the choice of the smoothing constant (α).

Equation (2.52) becomes invalid (does not provide accurate forecasting) for cases where data contain significant trend. Therefore, other exponential smoothing models must be used to include the effect of the trend. Consequently, we introduce the *double-exponential smoothing method* to correct for trend.

2.3.5. DOUBLE EXPONENTIAL SMOOTHING

To make a forecast using the double-exponential smoothing method, we assume that the data do not contain seasonal, cyclic, or irregular variations. The data show only a linear trend (the following procedure can be extended to data with a nonlinear trend). Eq. (2.53) represents the data generation process.

$$f_t = a_0 + a_1 t + \epsilon_t \tag{2.53}$$

where a_0 and a_1 are the parameters of the process and ϵ has an expected value of 0 and a variance σ_ϵ^2. Recalling Eq. (2.52) and letting $\beta = 1 - \alpha$, we have

$$h_t = \alpha f_t + \alpha\beta f_{t-1} + \alpha\beta^2 f_{t-2} + \cdots + \alpha\beta^{t-1} f_1 + \beta^t h_0 \tag{2.54}$$

Eq. (2.54) can be rewritten as

$$h_t = \alpha \sum_{i=0}^{t-1} \beta^i f_{t-i} + \beta^t h_0$$

The expectation of h_t is

$$E[h_t] = \alpha \sum_{i=0}^{t-1} \beta^i E[f_{t-i}] + \beta^t h_0$$
$$= \alpha \sum_{i=0}^{t-1} \beta^i [a_0 + a_1(t-i)] + \beta^t h_0$$

As time periods increase ($t \to \infty$), $\beta^t \to 0$ and

$$E[h_t] = a_0 + a_1 t - \frac{\beta}{\alpha} a_1$$

or

$$E[h_t] = E[f_t] - \frac{\beta}{\alpha} a_1 \qquad (2.55)$$

Equation (2.55) implies that the simple exponential smoothing lags the actual demand by $(\beta/\alpha)a_1$. The double-exponential smoothing is obtained by considering h_t in Eq. (2.51) to be the actual demand at time t (i.e., h_t becomes f_t). Consequently,

$$h_{2,t} = \alpha h_t + \beta h_{2,t-1} \qquad (2.56)$$

where $h_{2,t}$ is the double-exponential smoothing forecast value. Following Eq. (2.55),

$$E[h_{2,t}] = E[h_t] - \frac{\beta}{\alpha} a_1 \qquad (2.57)$$

or

$$a_1 = \frac{\alpha}{\beta} (E[h_t] - E[h_{2,t}]) \qquad (2.58)$$

From Eqs. (2.55) and (2.57) we obtain

$$E[f(t)] = 2E[h_t] - E[h_{2,t}] \qquad (2.59)$$

The estimates of a_0 and a_1 can be obtained from Eqs. (2.59) and (2.58), respectively, as

$$\hat{a}_0(t) = 2h_t - h_{2,t} \qquad (2.60)$$

$$\hat{a}_1(t) = \frac{\alpha}{\beta} (h_t - h_{2,t}) \qquad (2.61)$$

Since h_t and $h_{2,t}$ always follow the actual data, we must calculate initial values for h_0 and $h_{2,0}$:

$$h_0 = \hat{a}_0(0) - \hat{a}_1(0) \frac{\beta}{\alpha} \qquad (2.62)$$

$$h_{2,0} = \hat{a}_0(0) - 2\hat{a}_1(0) \frac{\beta}{\alpha} \qquad (2.63)$$

It should be noted that $\hat{a}_1(0)$ is an initial estimate of the slope at $t = 0$, whereas $\hat{a}_0(0)$ is the intercept of the linear model at $t = 0$, or simply, it is the estimate of h_0.

Table 2-20. DATA FROM EXAMPLE 2-1

t	0	1	2	3	4
$f(t)$	100	115	116	125	135

EXAMPLE 2-10

Use the double-exponential smoothing method to determine next year's forecast for the data of Example 2-1, shown in Table 2-20. Assume that $\alpha = 0.3$.

SOLUTION

To determine h_0 and $h_{2,0}$, we first need to estimate $\hat{a}_0(0)$ and $\hat{a}_1(0)$. We can show that $\hat{a}_0(0) = 100$ and $\hat{a}_1(0) = 8$. These values can be estimated as

$$\hat{a}_0(0) = f(0) = 100$$

$$\hat{a}_1(0) = \frac{f_4 - f_0}{4} = \frac{35}{4} = 8.75$$

We choose $\hat{a}_1(0) = 8$ as that of Example 2-2. We utilize Eqs. (2.62) and (2.63) to get h_0 and $h_{2,0}$:

$$h_0 = 100 - 8\left(\frac{0.7}{0.3}\right) = 81.33$$

$$h_{2,0} = 100 - 2(8)\left(\frac{0.7}{0.3}\right) = 62.66$$

Also, h_t and $h_{2,t}$ are estimated by using Eqs. (2.51) and (2.56). Their calculations are shown in Table 2-21. The forecast for $t = 5$ is obtained by considering $t = 4$ as the initial point for following forecasts, [i.e., we need to estimate $\hat{a}_0(4)$ and $\hat{a}_1(4)$]. These can be obtained from Eqs. (2.60) and (2.61).

$$\hat{a}_0(4) = 2h_4 - h_{2,4} = 2(115.16) - 95.91 = 134.4$$

$$\hat{a}_1(4) = \frac{0.3}{0.7}(115.16 - 95.91) = 8.25$$

Table 2-21. CALCULATIONS FOR DOUBLE-EXPONENTIAL SMOOTHING

t	f_t	h_t	$h_{2,t}$
0	100	81.33	62.66
1	115	91.43	71.29
2	116	98.80	79.54
3	125	106.66	87.67
4	135	115.16	95.91

Table 2-22. MEAN ABSOLUTE ERRORS FOR A SPECIFIED RANGE OF α

α	α_{min}	$\alpha_{min} + \Delta$	$(\alpha_{min} + 2\Delta)$	$\cdots$	α_{max}
Mean Absolute Deviation	$\dfrac{\sum_i e_{1i}}{n}$	$\dfrac{\sum_i e_{2i}}{n}$	$\dfrac{\sum_i e_{3i}}{n}$	$\cdots$	$\dfrac{\sum_i e_{mi}}{n}$

The forecasting model becomes

$$h_{4+\tau} = \hat{a}_0(4) + \hat{a}_1(4)(\tau)$$

where τ is the number of time periods where forecasting is desired. Next year's forecasted demand is

$$h_5 = 134.4 + (8.25)(1) = 142.65 \tag{2.64}$$

or

next year's demand = 142,650 units

Comparison of the forecast from the linear regression model (142,200 units) with that of the double-exponential smoothing model (142,650 units) indicates that the two approaches yield very close results. This statement is true for data that exhibit a linear trend. However, when the data show a nonlinear trend, higher-order exponential smoothing should be developed by following the same steps for the double-exponential smoothing (or second-order exponential smoothing) model.

2.3.6. CHOICE OF α

Brown [2] suggested a practical range of α between 0.01 and 0.3. A small value of α should be selected for noisy data. If the data are stable, larger values of α can be used. The *optimum* value of α can be obtained by using an *iterative* approach, which cycles through the data with a minimum α value and the absolute errors (e_i) between the forecasted values and their respective actual values are then averaged. Then α is incremented by $\Delta\alpha$ and the process is repeated until all values of α (m values) within a specified range are exhausted. The optimum value of α^* is that corresponding to minimum average of the absolute errors as shown in Table 2-22. These averages of absolute errors are referred to as mean absolute deviations (MAD). The following example illustrates the foregoing procedure for $n = 5$.

EXAMPLE 2-11

Find the optimum value of α for the data given in Example 2-1. Use the simple exponential smoothing method with $0.1 \leq \alpha \leq 0.4$. Calculations are shown in Tables 2-23 through 2-26, and a summary of MAD values in Table 2-27.

Table 2-23. CALCULATIONS FOR
$\alpha = 0.1$

t	f_t	h_t	$|f_t - h_t|$
0	100	118.2	18.20
1	115	117.88	2.88
2	116	117.69	1.69
3	125	118.42	6.58
4	135	120.08	14.91

Table 2-24. CALCULATIONS FOR
$\alpha = 0.2$

t	f_t	h_t	$|f_t - h_t|$
0	100	118.20	18.20
1	115	117.56	2.56
2	116	117.24	1.24
3	125	118.79	6.21
4	135	122.03	12.97

SOLUTION

Iteration 1: Choose $\alpha = 0.1$, $h_0 = (\sum_{t=0}^{4} f_t)/5 = 118.2$.
Mean absolute deviation (MAD):

$$\text{MAD}_1 = \frac{\sum_{t=1}^{n} |f_t - h_t|}{n}$$

$$\text{MAD}_1 = \frac{44.26}{5} = 8.852$$

Iteration 2: Set $\alpha = 0.2$.

$$\text{MAD}_2 = \frac{41.18}{5} = 8.236$$

Iteration 3: $\alpha = 0.3$.

$$\text{MAD}_3 = 7.598$$

Iteration 4: $\alpha = 0.4$

$$\text{MAD}_4 = 7.086$$

Table 2-25. CALCULATIONS FOR
$\alpha = 0.3$

t	f_t	h_t	$|f_t - h_t|$
0	100	118.20	18.20
1	115	117.24	2.24
2	116	116.86	0.86
3	125	119.30	5.70
4	135	124.01	10.99

Table 2-26. CALCULATIONS FOR
$\alpha = 0.4$

t	f_t	h_t	$|f_t - h_t|$
0	100	118.60	18.60
1	115	117.16	2.16
2	116	116.99	0.69
3	125	120.01	4.99
4	135	126.01	8.99

Table 2-27. SUMMARY
OF MAD

α_i	MAD_i
0.1	8.852
0.2	8.236
0.3	7.598
0.4	7.086*

*$\alpha = 0.4$.

The standard deviation of a forecast.   As in the case of linear regression, it is usually desirable to compute a confidence interval around the forecast mean. For exponential smoothing methods, it is usual to estimate the standard deviation of a forecast using historical forecast errors. For m historical observations, the statistic is computed:

$$\hat{\sigma}_f = \sqrt{\sum_{j=1}^{m} (e_j - \bar{e})^2/(m - 1)}$$

where $\hat{\sigma}_f$ = estimated standard deviation of forecast errors over a
forecast horizon of one period
e_j = historical forecast error in period j $(j = 1, \ldots, m)$
$\bar{e}$ = mean forecast error; $\bar{e} = \sum_{j=1}^{m} [f(t) - h(t)]/m$

If errors are normally distributed, $\bar{e} \simeq 0$.

An alternative statistic to calculate in place of the standard deviation is the *mean absolute deviation* (MAD). This statistic is defined as follows:

$$\text{MAD} = \frac{\sum_{j=1}^{m} |e_j|}{m}$$

where MAD = mean absolute deviation of forecast errors over a
forecast horizon of one period
$|e_j|$ = absolute value of the forecast error in period j

As in the case of exponentially weighting forecast equation parameters, greater weight can be applied to current estimates of MAD through the recursion

$$\text{MAD}_t = \alpha(f(t) - h(t)) + (1 - \alpha)\text{MAD}_{t-1}$$

Finally, it can be shown that when forecast errors are normally distributed, the following simple conversion can be used to compute σ_f.

$$\sigma_f = (1.25)\text{MAD}$$

2.3.7 WINTERS' METHOD FOR SEASONAL VARIATION

When seasonal variation is present in the demand data, a popular choice of exponential smoothing model is Winters' method [16]. Winters' method assumes three components to the model: a *permanent* component, a *trend*, and a *seasonal* component. Each component is continuously updated using a smoothing constant applied to the most recent observation and the last estimate.

Winters' model assumes a data-generating process of the form

$$f_t = (a_{0,t} + a_{1,t}t)C_t + \epsilon_t$$

where C_t is the seasonal factor. The parameters $a_{0,t}$, $a_{1,t}$, and C_t are continuously updated using an exponential smoothing procedure.

$$a_{0,t} = \alpha \frac{f_t}{C_{t-N}} + (1 - \alpha)(a_{0,t-1} + a_{1,t-1})$$

where $a_{0,t}$ = exponentially smoothed level of the process at the end of period t

f_t = actual sales of period t

N = number of periods in the season

$a_{1,t-1}$ = trend for period $t - 1$

α = smoothing constant for a_0

Similarly, for the seasonal factor:

$$C_t = \gamma \frac{f_t}{a_{0,t}} + (1 - \gamma)C_{t-N}$$

where γ is the smoothing constant for C_t. For updating the trend,

$$a_{1,t} = \phi(a_{0,t} - a_{0,t-1}) + (1 - \phi)a_{1,t-1}$$

where ϕ is the smoothing constant for a_1. The model is then constructed from the most recent updated parameters as follows:

$$h_t = (a_{0,t} + a_{1,t})C_t$$

As in any exponential smoothing model, in order to begin the forecasting process, a set of initial conditions are required for the parameters. There are several ways to accomplish this: one method is as follows.

—*Slope:* Compute the slope from the average sales of the first two years of data.

$$\bar{S}_1 = \frac{\sum\limits_{t=1}^{N} f_t}{N}$$

$$\bar{S}_2 = \frac{\sum\limits_{t=N+1}^{2N} f_t}{N}$$

$$a_1 = \frac{\bar{S}_2 - \bar{S}_1}{N}$$

—*Level:* Compute the level at the end of period $2N$ using $\bar{S}_2$ and the computed slope, a_1.

$$a_{0,\,2N} = \bar{S}_2 + a_1 \frac{N-1}{2}$$

—*Seasonal:* Compute the initial seasonal factors as the deviation of actuals from the function with slope a_1 and intercept a_0.

$$C_t = \frac{f_t}{a_0 + a_1 t}$$

where $a_0 = a_{0,\,2N} - (2N)a_1$. The 2 years of seasonal factors are then averaged to yield a single set of seasonal factors. These must then be normalized:

$$\frac{\sum\limits_{t=1}^{N} C_t}{N} = 1$$

EXAMPLE 2-12

Given the data for the sale of snow tires in Table 2-28, forecast the sales for the first 3 months of year 3.

SOLUTION

The parameters of the model can be initialized as follows:

Slope:

$$\bar{S}_1 = \frac{\sum\limits_{t=1}^{N} f(t)}{N} = \frac{2522}{12} = 210.2$$

$$\bar{S}_2 = \frac{\sum\limits_{t=N+1}^{2N} f(t)}{N} = \frac{2817}{12} = 234.8$$

Table 2-28. SALES OF SNOW TIRES

	Number of Tires Sold	
Month	Year 1	Year 2
Jan.	195	229
Feb.	252	271
Mar.	299	333
Apr.	276	305
May	258	277
June	210	239
July	192	211
Aug.	154	193
Sept.	141	155
Oct.	128	159
Nov.	200	209
Dec.	217	236
	2522	2817

$$a_1 = \frac{\bar{S}_2 - \bar{S}_1}{N} = \frac{234.8 - 210.2}{12} = 2.05$$

Level:

$$a_{0,\,2N} = \bar{S}_2 + a_1 \frac{N-1}{2} = 234.8 + 2.05\left(\frac{11}{2}\right) = 246.08$$

Seasonal:

$$a_0 = a_{0,\,2N} - (2N)a_1 = 246.08 - (24)(2.05) = 196.88$$

$$C_t = \frac{f_t}{a_0 + a_1 t}$$

t (year 1)	$a_0 + a_1 t$	C_t	t (year 2)	$a_0 + a_1 t$	C_t	$\bar{C}_t$
1	198.93	0.98	13	223.53	1.02	1.00
2	200.98	1.25	14	225.58	1.20	1.23
3	203.03	1.47	15	227.63	1.46	1.47
4	205.08	1.35	16	229.68	1.33	1.34
5	207.13	1.25	17	231.73	1.20	1.23
6	209.18	1.00	18	233.78	1.02	1.01
7	211.23	0.91	19	235.83	0.89	0.90
8	213.28	0.72	20	237.88	0.81	0.77
9	215.33	0.65	21	239.93	0.65	0.65
10	217.38	0.59	22	241.98	0.66	0.63
11	219.43	0.91	23	244.03	0.86	0.88
12	221.48	0.98	24	246.08	0.96	0.97
						12.08

After normalization, seasonal indexes are

t	Normalized C_t
1	0.99
2	1.22
3	1.46
4	1.33
5	1.22
6	1.00
7	0.89
8	0.76
9	0.64
10	0.62
11	0.87
12	0.96
	12.00

The forecast for the first 3 months of year 3 is as follows:

$$h(\text{Jan.}) = [246.08 + (2.05)(1)](0.99) = 245.64$$

$$h(\text{Feb.}) = [246.08 + (2.05)(2)](1.22) = 305.21$$

$$h(\text{Mar.}) = [246.08 + (2.05)(3)](1.46) = 368.25$$

Parameter updating is usually done when a new actual result for a period is known. For example, if the actual sales for January were 245 and

$$\alpha = \gamma = \phi = 0.3$$

the updating would proceed as follows:

$$a_0 = 0.3\left(\frac{245}{0.99}\right) + (0.7)(246.08 + 2.05) = 247.93$$

$$C_t = 0.3\left(\frac{245}{247.93}\right) + (0.7)(0.99) = 0.97$$

$$a_1 = 0.3(247.93 - 246.08) + (0.7)(2.05) = 1.99$$

Forecasts for February, March, and April are then made as follows:

$$h(\text{Feb.}) = [247.93 + (1.99)(1)](1.22) = 304.90$$

$$h(\text{Mar.}) = [247.93 + (1.99)(2)](1.46) = 367.78$$

$$h(\text{Apr.}) = [247.93 + (1.99)(3)](1.33) = 337.68$$

2.4. Comparison of Time-Series Forecasts

So far we have presented several techniques for forecasting future demands or sales. One might ask: "Which technique should I use to obtain the *best* fit for the data?"

As we indicated earlier, the first step in forecasting is plotting and observing the data for trend, cyclical, seasonal, and irregular variations. An appropriate model should then be chosen to fit the data. If there is more than one possible model, a unified and suitable measure of the accuracy of the forecast should be chosen such as MAD, coefficient of correlation, coefficient of determination, or standard error of estimate.

PROBLEMS

2-1. Listed below are the sales for the past 10 years of a small manufacturing company.

Year	Sales
1	$100,000
2	98,000
3	95,000
4	99,000
5	110,000
6	117,000
7	129,000
8	140,000
9	152,000
10	157,000

(a) Plot the sales and draw a trend line by inspection.
(b) Use Gregory–Newton interpolation formulas to predict sales for year 11.
(c) Compare the forecasts from parts (a) and (b) and evaluate them.

Table 2-29.

Year	Demand (10^3 units)
1	50.7
2	55.4
3	59.6
4	61.0
5	58.0
6	60.5
7	66.0
8	70.5
9	77.8
10	87.6
11	94.8
12	100.7

Table 2-30.

Year	Production Quantity (10^6 units)
1	6.0
2	10.0
3	13.5
4	16.5
5	19.0
6	21.3
7	23.4
8	25.4
9	27.2
10	29.0

2-2. The manager of a warehouse distribution center would like to determine the demand for items in the center in order to decide on the size of a new layout. The demand is shown in Table 2-29.

(a) Draw a trend line by eye for these data.
(b) Use a simple regression model to forecast the demand for year 13.
(c) Calculate the 95% confidence interval around the forecast.
(d) Calculate the correlation coefficient.

2-3. The forecast model for production quantities of an item can be expressed by

$$y = \frac{x}{\alpha_0 x - \alpha_1}$$

where y = production quantity
x = time period, years
α_0, α_1 = constants

(a) Using Table 2-30, determine the production quantity for next year.
(b) Determine the coefficient of determination.

2-4. The seasonal demands for a product are shown in Table 2-31.

Table 2-31.

Year	Month	Demand (units)
1	Jan.	100
	Feb.	120
	Mar.	140
	Apr.	160
	May	155
	June	150
	July	145
	Aug.	140
	Sept.	135
	Oct.	145
	Nov.	160
	Dec.	200
2	Jan.	210
	Feb.	230
	Mar.	250

Table 2-32.

Productivity, P (%)	Buffer Size, Z	ρ
40	2	0.90
50	3	0.90
55	3	0.75
60	4	0.80
65	8	0.70
68	7	0.65
70	9	0.60
75	15	0.60
80	20	0.50
85	38	0.51
90	60	0.48
92	75	0.45
93	85	0.44
94	110	0.43
95	140	0.43

(a) Plot the points and draw a trend line by eye.
(b) Use a trigonometric function and fit the data by the least-squares method.
(c) Determine the coefficient of determination.
(d) What are the forecasted demands for the following 9 months?

2-5. As shown in Table 2-32, the productivity of a transfer line is a function of the size of buffer storage and the ratio of the failure and repair rates of the production machines (ρ).

(a) Estimate the productivity P for a production system with a buffer storage Z and a failure to repair rate ratio ρ, by utilizing the multiple linear regression technique. Find P if $Z = 180$ and $\rho = 0.35$.
(b) Calculate the 95% confidence interval around the forecast.
(c) Determine the correlation coefficient.

2-6. The production quantity of product A is dependent on the sales of two other products, B and C. The sales of these products are given in Table 2-33.

Table 2-33. NUMBER OF UNITS SOLD FROM PRODUCT

A	B	C
100	80	50
200	90	30
300	70	40
400	60	60
500	50	80
600	95	35
700	45	75

Table 2-34.

Month	First Year	Second Year
Jan.	29	50
Feb.	57	60
Mar.	62	64
Apr.	83	87
May	104	109
June	121	127
July	138	139
Aug.	165	169
Sept.	71	65
Oct.	52	55
Nov.	67	57
Dec.	62	50

(a) What are the expected sales of product A if 40 and 50 units are expected to be sold from products B and C, respectively?

(b) What is the correlation coefficient?

2-7. Use a 3-month moving average and a 5-month moving average to forecast demand for the first 6 months of year 2 for data given in Prob. 2-4. Compare the forecasted values.

2-8. Use a 3-month moving total and a 5-month moving total to forecast demand for the data given in Prob. 2-4. Compare results with those of Prob. 2-7.

2-9. Records show that a company has the demand shown in Table 2-34 for one of its products.

(a) Establish a monthly forecast for the third year by using the moving-average approach.

(b) Establish a monthly forecast for the first 6 months of the third year by using exponential smoothing with $\alpha = 0.25$; compare the forecast of (a) and (b).

(c) Calculate the 95% confidence interval around the forecast.

2-10. Use the double-exponential smoothing method to determine the forecast of January of the third year for the data given in Prob. 2-9. Assume that $\alpha = 0.33$.

2-11. A 2-year sales history by quarter for a seasonal product is given below.

	Sales	
Quarter	First Year	Second Year
1	30	42
2	48	58
3	60	74
4	35	44

(a) Using Winters' method, forecast the quarterly sales for the third year.

(b) Assume that the actual sales for the first quarter of year 3 is 48 units. What is the revised forecast for the second quarter of the third year? Use $\alpha = \gamma = \phi = .2$.

(c) Assume the following actual sales occurred during the third year.

Quarter	Sales
1	48
2	70
3	95
4	50

What is the 95% confidence interval around a one period forecast?

REFERENCES

[1] BEDWORTH, DAVID D., *Industrial Systems.* New York, N.Y.: Ronald Press Co., 1973.

[2] BROWN, ROBERT G., *Forecasting and Prediction of Discrete Time Series.* Englewood Cliffs, N.J.: Prentice-Hall Inc., 1963.

[3] ———, *Statistical Forecasting for Inventory Control.* New York, N.Y.: McGraw-Hill Book Co., 1959.

[4] ———, *Decision Rules for Inventory Management.* New York, N.Y.: Holt, Rinehart & Winston, 1967.

[5] CHATTERJEE, SAMPRIT, AND BERTRAM PRICE, *Regression Analysis by Example.* New York, N.Y.: John Wiley & Sons, 1977.

[6] CHOW, WEN M., "Adaptive Control of the Exponential Smoothing Constant," *Journal of Industrial Engineering,* XVI, no. 5 (1965) pp. 314–317.

[7] DRAPER, NORMAN R., AND HARRY SMITH, *Applied Regression Analysis.* New York, N.Y.: John Wiley & Sons, 1968.

[8] EILON, SAMUEL, *Elements of Production Planning and Control.* Chapter 6, New York, N.Y.: Macmillan Co., 1962.

[9] GIFFIN, WARREN C., *Introduction to Operations Engineering.* Homewood, Ill.: Richard D. Irwin, 1971.

[10] HORNBECK, ROBERT W., *Numerical Methods.* New York, N.Y.: Quantum Publishers, Inc., 1975.

[11] MONTGOMERY, DOUGLAS C., "Introduction to Short-Term Forecasting," *Journal of Industrial Engineering,* XIX, no. 10 (1968) pp. 500–503.

[12] ———, "Adaptive Control of Exponential Smoothing Parameters by Evolutionary Operation," *AIIE Transactions,* 2, no. 3 (1970) pp. 268–269.

[13] SHORE, BARRY, *Operations Management.* New York, N.Y.: MacGraw-Hill Book Co., 1973.

[14] SWEET, ARNOLD L., "Adaptive Smoothing for Forecasting Seasonal Series," *AIIE Transactions*, Vol. 13, no. 3, Sept., 1981.

[15] TAYLOR, SAM G., "Initialization of Exponential Smoothing Forecasts," *AIIE Transactions*, Vol. 13, no. 3, Sept., 1981.

[16] WINTERS, PETER R., "Forecasting Sales by Exponentially Weighted Moving Averages," *Management Science*, 6, no. 3 (1960) pp. 324–342.

INVENTORY SYSTEMS

3.1. Introduction

The demand forecast, discussed in Chapter 2, is one of the primary inputs for making inventory decisions. We define *inventory* as the raw material, semi-finished parts and assemblies, and finished goods that are in the production system at any point in time. Inventories serve as a buffer between stages of the production system and between the production system and its customers.

The main objective of the analysis of an inventory system is to find the answers to the following two questions:

1. How much should be ordered (or produced)?
2. When should the orders be placed such that the total inventory costs are minimized?

Inventory costs can be classified as (1) the cost of carrying inventories (*holding cost*), (2) the cost of incurring shortages (*opportunity cost*), and (3) the cost of replenishing inventories (*order cost*). These costs are now explained in detail.

3

3.2. Inventory Costs

3.2.1. Inventory Carrying Costs

The cost of carrying inventory can be broken down into several components: (1) *the opportunity cost* of money being tied up in inventory, for example, the interest foregone because that money is not placed in an interest-bearing account; (2) *storage and space charges*, representing the cost of providing storage space as well as its cost of maintenance (it also represents the cost of handling units and the cost of computers and peripherals which are used to keep track of the inventory); (3) *taxes and insurance* and the cost of *physical deterioration and its prevention*, as in the case of inventorying dry cell batteries, vegetables, dairy products, and some ceramic and electronic products; and (4) the *cost of obsolescence* due to technological change, as in the fields of personal computers, robotics, and communications equipment.

3.2.2. Shortage Costs

This cost is incurred if units of inventory are not available when demanded. It is the cost of lost sales, of loss of goodwill, of overtime payments, of customer dissatisfaction, and of special administrative efforts (telephone calls, memos, etc.) resulting from the inability to meet demand. There are two types of shortage costs: (1) one-time shortage cost per unit short, independent of the duration of the shortage; and (2) shortage cost per unit short per unit time.

3.2.3. ORDERING COST

Ordering costs include the cost of preparing and placing orders for replenishing inventories, the cost of handling and shipments of orders, the cost of machine setups for production runs, the cost of inspection of received orders in inventory, and all costs that do not vary with the size of the order. Sometimes it is difficult to determine these costs in details; therefore, the analyst may combine costs. For the inventory models developed in this text, it is assumed that the ordering cost does not vary with the size of the order.

3.3. The Terminology of Inventory Systems

The following definitions are used in conjunction with the analysis of inventories.

3.3.1. DEMAND

Inventory decisions (policies, quantities to be ordered, etc.) are made with reference to future demand. The *demand* can be deterministic or probabilistic and static or dynamic in nature. It is the usage rate of a product. The demand rate is the quantity demanded per unit time. This demand rate can be uniform, infinite, power, or others. In general, the amount in inventory at time T, $Q(T)$, is given by

$$Q(T) = Q_0 - X \sqrt[n]{\frac{T}{t}} \tag{3.1}$$

where $Q(T)$ = amount in inventory at time T
   Q_0 = amount in inventory at the beginning of the period
         (at $T = 0$)
   X = demand size during the period t
   t = length of the period
   n = demand-pattern index

The demand patterns as given in Naddor [10] for different values of n is shown in Fig. 3-1. When the demand is described by a probability distribution, the demand rate is estimated as the expected value of the demand rate.

3.3.2. LEAD TIME AND REPLENISHMENT RATE

Lead time is the time interval between the time when an order is placed and the time when it is actually received in the inventory. The lead time can be deterministic or probabilistic and constant or time varying. The *replenishment*

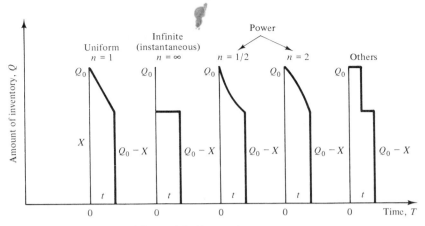

Figure 3-1. Demand patterns.

rate (procurement rate, production rate) is the rate at which the inventory builds up. Figure 3-2 shows different replenishment patterns.

3.3.3. REORDER LEVEL

The *reorder level* is the inventory level at which orders are placed for replenishing the inventory. The reorder level is a function of the lead time demand.

3.3.4. SAFETY STOCK

Safety stock is inventory that is carried to prevent a stockout when there is uncertainty in the demand or supply process. Safety stock is a function of the lead time demand.

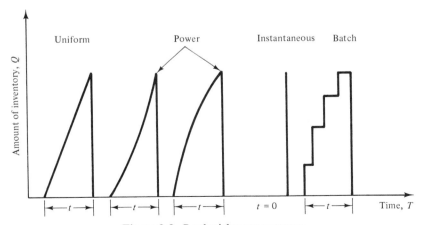

Figure 3-2. Replenishment patterns.

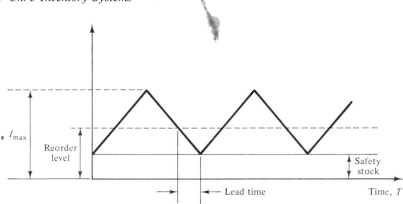

Figure 3-3. Typical inventory system.

In a typical inventory system orders are received at a replenishment rate until the total order quantity is received. At this time the inventory level reaches its maximum value. The demand for the items will cause the inventory level to decrease at a rate equal to the demand rate until the inventory level reaches the reorder level, at which time another order is placed. Figure 3-3 represents the pattern of on-hand inventory in a typical inventory system.

3.4. Inventory Policies

Inventory policies refer to the review and ordering discipline used in *controlling the inventory* (when orders should be placed and how much). The most commonly used inventory policies are presented here.

3.4.1. PERIODIC-REVIEW POLICY

Under this policy, inventory levels are observed at equal intervals of time, T. We refer to T as the review period length. If at the end of period T, the inventory level is higher than a predetermined reorder level, no action is taken. However, if it is less than or equal to the reorder level, an order is placed to bring the inventory to the target (maximum) level.

I_i = inventory level at the end of period T_i
r = reorder level
I_{max} = maximum acceptable inventory level (target level)
Q_i = order size at period i (difference between I_{max} and I_i)

This policy can be presented as

$$Q_i = \begin{cases} 0 & \text{if } I_i > r \\ I_{max} - I_i & \text{if } I_i \le r \end{cases}$$

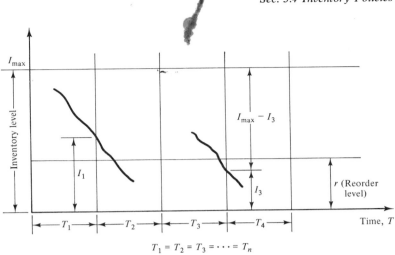

Figure 3-4. Periodic-review policy.

There are three basic parameters that are needed to define this policy: I_{max}, r, and T. Therefore, the optimal values for I_{max}, r, and T must be determined such that the total inventory cost of this policy is minimized. This policy is presented graphically in Fig. 3-4. From this figure it is clear that no orders need to be placed at the end of period T_1, while an order of $I_{max} - I_3$ must be placed at the end of period T_3.

3.4.2. ORDER UP TO I_{max} POLICY

This is a special case of the (I_{max}, r, T) periodic review policy. Under this policy, the reorder level r is set to equal I_{max}. Consequently, an order of size $Q_i = I_{max} - I_i$ is always placed at the end of the period T_i. I_{max} and T are the only two parameters needed to define this policy.

3.4.3. CONTINUOUS-REVIEW POLICY

Under this policy, the inventory level is continuously monitored and orders of size $Q_i = I_{max} - I_i$ are always placed if the inventory level I_i drops to the reorder level r, or below. The difference between the periodic review policy and the continuous review policy is that under the former policy, orders *may or may not* be placed at the end of period T_i depending on the inventory level, whereas under the latter policy, orders are *always* placed when the inventory level drops to, or below, r (independent of the length of the time period). Figure 3-5 represents this policy.

3.4.4. FIXED-REORDER QUANTITY POLICY

This policy is similar to the continuous review policy with the exception that the units are taken out from the inventory one at a time. Therefore, the inventory level can be observed when it drops to exactly r. Consequently, a

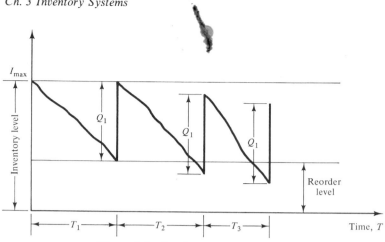

Figure 3-5. Continuous-review policy.

fixed order of size Q is always placed when $I_i = r$. Q and r are the only parameters needed to define this policy; it is denoted by (Q, r).

3.4.5. BASE STOCK POLICY

Under this policy we set the reorder level $r = I_{max}$ and orders are placed after each withdrawal from the inventory. Consequently, the sum of the amount of inventory on hand, I_i, and the amount being ordered, Q_i, must equal I_{max} at all times. The maximum inventory level, I_{max}, is referred to as the *base stock level*.

3.5. Demand Characteristics and Inventory Models

Demand and lead time are the most common sources of uncertainty in any inventory system. We can classify inventory models with respect to the nature of the demand as follows:

1. *Static deterministic inventory models:* In these models, the demand is deterministic in nature (total number of units demanded over a fixed period of time is known and constant), and the demand rate is the same for every period.
2. *Dynamic deterministic inventory models:* The demand for each period is known and constant, but the demand rate may vary from one period to the other.
3. *Static probabilistic inventory models:* The demand is a random variable, having a probability distribution that depends on the length of

the period. The probability distribution of the demand is the same for each period.

4. *Dynamic probabilistic inventory models:* These models are similar to model 3 with the exception that the probability distribution of the demand may vary from one period to the other.

3.6. Analysis of Deterministic Inventory Models

In this section algebraic analysis of deterministic inventory systems are presented. The following notations will be used:

C = purchase price per unit (if purchased) or the unit variable cost of production (if produced)

D = demand rate, units per year

A = fixed cost of a replenishment order (order cost) or setup cost of production

P = production or replenishment rate, units per year $(P > D)$

h = inventory carrying cost per unit per year ($/unit/year), usually expressed as $h = iC$, where i is the annual inventory carrying cost rate

I_{max} = maximum on-hand inventory level, units

I = average on-hand inventory level, units

S_{max} = maximum shortage permitted, units

S = average shortage, units

r = reorder level, units

Q = order quantity, units

W = shortage cost per unit short, independent of the duration of the shortage

W_1 = shortage cost per unit short per year

T = cycle length, the length of time between production runs

TC = total annual cost, which is a function of the inventory policy

L = lead time, the length of time between placement and receipt of orders

3.6.1 SINGLE-PRODUCT MODEL

In this model we consider an inventory system with a constant demand rate D. The production rate P is finite (i.e., units produced are added to the inventory one at a time). The objectives of the analysis are to determine the optimal quantity Q^* to be ordered and the optimal shortage S^*_{max} allowed such that the total annual cost of the inventory system is minimized. Since the decision variables of this system are S^*_{max} and Q^*, the total-cost equation must be expressed in terms of S_{max} and Q. Other models will be presented later for which the total-cost equation is expressed in terms of other decision variables.

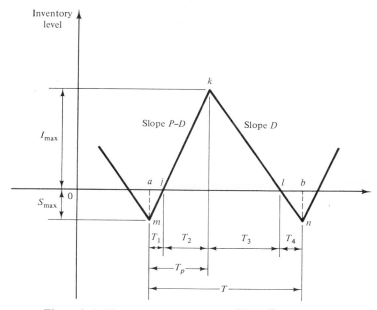

Figure 3-6. Single-product model with constant demand.

Figure 3-6 shows the change in the inventory level within a cycle T. When production starts (or orders start to arrive into the inventory) at point a, the inventory level will increase at a rate $P - D$, satisfying the backorders first and the current demand, until Q units are added to the inventory (point k). The inventory level will decrease at a rate D during the time period $T_3 + T_4$ and another cycle starts at point b, where the inventory level will increase at a rate $P - D$ until the quantity Q is produced, and so on. From Fig. 3-6 we obtain:

Time to produce a lot Q:

$$T_P = Q/P \tag{3.2}$$

Maximum inventory level:

$$I_{max} = [Q/P] \, (P - D) - S_{max}$$

or

$$I_{max} = Q\left(1 - \frac{D}{P}\right) - S_{max} \tag{3.3}$$

The total-cost equation consists of several cost components. The first component is the cost of placing an order (or setup cost). This has already been expressed as the fixed amount A.

The second component of the average cycle cost is the cost of carrying the inventory for one cycle. This cost is incurred during time periods T_2 and T_3 (these are the only time periods where we have inventory on hand during the total cycle, T).

$$T_2 = \frac{I_{max}}{P - D} \qquad \text{(time to build the inventory from zero to } I_{max})$$

$$T_3 = \frac{I_{max}}{D} \qquad \text{(time to consume the amount in inventory from } I_{max} \text{ to zero)}$$

The average inventory I over the cycle T is the area of the triangle jkl divided by T.

$$I = \frac{1}{2T}(T_2 + T_3)I_{max}$$

Substituting $T = Q/D$ and T_2, T_3, and I_{max} as given above, we obtain

$$I = \frac{[Q(1 - D/P) - S_{max}]^2}{2Q(1 - D/P)} \tag{3.4}$$

The average carrying cost over the cycle T is hTI.

The third component of the inventory cost is the shortage cost, which occurs over the time T_1 and T_4 during the cycle T.

$$T_1 = \frac{S_{max}}{P - D} \qquad \text{(time to eliminate the backorder)}$$

$$T_4 = \frac{S_{max}}{D} \qquad \text{(time to build a backorder of } S_{max})$$

The average shortage S over the cycle T is the sum of the areas of the two triangles amj and lbn divided by T.

$$S = \frac{1}{T}\frac{S_{max}^2}{2D(1 - D/P)} = \frac{S_{max}^2}{2Q(1 - D/P)} \tag{3.5}$$

and the average shortage cost during the cycle T is

$$W_1 TS + W S_{max}$$

The average cost per cycle is the sum of the ordering cost, item cost, carrying cost, and shortage cost

$$A + CQ + hTI + W_1 TS + W S_{max} \tag{3.6}$$

The total annual cost is obtained by multiplying Eq. (3.6) by the number of orders per year, D/Q. By substituting $h = iC$, we obtain

$$TC(Q, S_{max}) = \frac{AD}{Q} + CD + iCI + W_1 S + \frac{WS_{max} D}{Q} \tag{3.7}$$

Substitute for I and S from Eqs. (3.4) and (3.5):

$$TC(Q, S_{max}) = \frac{AD}{Q} + CD + \frac{iC[Q(1 - D/P) - S_{max}]^2}{2Q(1 - D/P)} \\ + \frac{W_1 S_{max}^2}{2Q(1 - D/P)} + \frac{WS_{max} D}{Q} \tag{3.8}$$

The decision variables of this model are Q and S_{max} and their optimum values can be obtained by solving the following simultaneous equations:

$$\frac{\partial TC(Q, S_{max})}{\partial Q} = 0 \tag{3.9}$$

$$\frac{\partial TC(Q, S_{max})}{\partial S_{max}} = 0 \tag{3.10}$$

The solution for $W_1 \neq 0$ is

$$Q^* = \sqrt{\frac{2AD}{iC(1 - D/P)} - \frac{(WD)^2}{iC(iC + W_1)}} \sqrt{\frac{iC + W_1}{W_1}} \tag{3.11}$$

$$S_{max}^* = \frac{(iCQ^* - WD)(1 - D/P)}{iC + W_1} \tag{3.12}$$

Special Cases: One of the most common inventory models occurs when the production rate P is infinite and shortages are not allowed. The total-cost equation for this model is obtained by letting $P \to \infty$ and $S_{max} \to 0$ in Eq. (3.8), which then reduces to

$$TC(Q) = \frac{AD}{Q} + CD + iC \frac{Q}{2} \tag{3.13}$$

and the optimal quantity Q^* is given by

$$Q^* = \sqrt{\frac{2AD}{iC}} \tag{3.14}$$

$h = iC$

Equation (3.14) is referred to as the EOQ (*economic order quantity*) model.

If the shortage cost per unit short is zero (i.e., $W = 0$), Eqs. (3.11) and (3.12) are reduced to

$$Q^*(W = 0) = \sqrt{\frac{2AD}{iC(1 - D/P)}} \sqrt{\frac{iC + W_1}{W_1}} \qquad (3.15)$$

$$S^*_{max}(W = 0) = \frac{iCQ^*(1 - D/P)}{iC + W_1} \qquad (3.16)$$

Substituting (3.15) and (3.16) into (3.8), respecting $W = 0$, yields

$$TC^*(Q, S_{max}) = CD + \sqrt{\frac{2ADiC(1 - D/P)W_1}{iC + W_1}} \qquad (3.17)$$

Another special case occurs when both W and W_1 are finite while the production rate P is infinite. The equations corresponding to this case are obtained by taking the limit of Eqs. (3.8), (3.11), and (3.12) as $P \to \infty$. This yields

$$TC(Q, S_{max}) = \frac{AD}{Q} + CD + \frac{iC(Q - S_{max})^2}{2Q} + \frac{S_{max}(W_1 S_{max} + 2WD)}{2Q} \qquad (3.18)$$

$$Q^* = \sqrt{\frac{2AD}{iC} - \frac{(WD)^2}{iC(iC + W_1)}} \sqrt{\frac{iC + W_1}{W_1}} \qquad (3.19)$$

$$S^*_{max} = \frac{iCQ^* - WD}{iC + W_1} \qquad (3.20)$$

The optimal value of Q^* for Eq. (3.15) is reduced to Eq. (3.21) when shortages are not allowed.

$$Q^* = \sqrt{\frac{2AD}{iC(1 - D/P)}} \qquad (3.21)$$

LIMITED PRODUCTION RATE "P"

EXAMPLE 3-1

A company XYZ purchases air filters which are used at the rate of 350 per year. The cost of each filter is $30 and the cost of placing each order is $10. The inventory carrying cost rate is 0.18. Shortage cost consists of two components:

1. Fixed cost of $0.30 per unit
2. Variable cost of $5 per unit short per year

Find both the optimal order quantity and the optimal shortage.

SOLUTION

Since the replenishment rate P is not given, it is assumed that $P \to \infty$.

$$D = 350 \text{ units/year}$$
$$i = 0.18$$
$$W = \$0.30$$
$$W_1 = \$5.0/\text{unit/year}$$
$$C = \$30.0/\text{unit}$$
$$A = \$10.0/\text{order}$$

Using Eqs. (3.19) and (3.20), we obtain

$$Q^* = \sqrt{\frac{2(10)(350)}{(0.18)(30)} - \frac{[(0.30)(350)]^2}{(0.18)(30)[(0.18)(30) + 5]}} \sqrt{\frac{(0.18)(30) + 5}{5}}$$

$$= 47.83 \simeq 48 \text{ units/year}$$

and

$$S^*_{max} = \frac{(0.18)(30)(48) - (0.30)(350)}{(0.18)(30) + 5}$$

$$= 14.82 \simeq 15 \text{ units}$$

The cycle length $T = (Q/D) = 0.137$ years $\simeq 50$ days. The total annual cost is calculated by using Eq. (3.22).

$$TC(Q^*, S^*_{max}) = \frac{AD}{Q^*} + CD + \frac{iC(Q^* - S^*_{max})^2}{2Q^*} + \frac{W_1 S^{*2}_{max}}{2Q^*} + \frac{W S^*_{max} D}{Q^*}$$

$$= \frac{10(350)}{48} + 30(350) + \frac{(0.18)(30)(48 - 15)^2}{2(48)} + \frac{5(15)^2}{2(48)} + \frac{(0.3)(350)(15)}{48} \quad (3.22)$$

$$= \$10,679$$

EXAMPLE 3-2

A metal-cutting center orders a spare part that is needed for its numerical controlled machines. The cost of the part is $60, the order cost is $20 per order, and the annual demand is 400 units per year. The carrying cost rate of the inventory is 0.24 per year. It is found that lack of spares costs the center $20 per spare short per year. What are the optimal order quantity and the optimal shortage level?

SOLUTION

$$D = 400 \text{ units/year}$$
$$C = \$60$$

$$A = \$20$$

$$i = 0.24$$

$$W_1 = \$20/\text{unit/year}$$

We assume that $P \to \infty$ and $W = 0$; then Eq. (3.11) becomes

$$Q^* = \sqrt{\frac{2AD}{iC}} \sqrt{\frac{iC + W_1}{W_1}}$$

$$= \sqrt{\frac{2(20)(400)}{(0.24)(60)}} \sqrt{\frac{(0.24)(60) + 20}{20}} \simeq 44 \text{ units}$$

$$S^*_{\max} = \frac{iCQ^*}{iC + W_1} = \frac{(0.24)(60)(44)}{(0.24)(60) + 20} \simeq 18 \text{ units}$$

$$T^* = \frac{44}{400} = 40 \text{ days}$$

and

$$TC(Q^*, S^*_{\max}) = \frac{AD}{Q^*} + CD + \frac{iC(Q^* - S^*_{\max})^2}{2Q} + \frac{W_1 S^{*2}_{\max}}{2Q^*}$$

$$= \frac{20(400)}{44} + 60(400) + \frac{(0.24)(60)(44 - 18)^2}{2(44)} + \frac{20(18)^2}{2(44)} = \$24,366$$

3.6.2. SENSITIVITY ANALYSIS

Suppose that in Example 3-1 the actual cost of processing an order was $20 and that we had erroneously measured it as $10. Therefore, the true Q^* and $S^*_{\max}$ should have been

$$Q = 71 \text{ units}$$

$$S^*_{\max} = 27 \text{ units}$$

and the total cost of the strategy, excluding CD term, that should have been adopted is

$$TC(Q^*, S^*_{\max}) = \frac{20(350)}{71} + \frac{(0.18)(30)(71 - 15)^2}{2(71)} + \frac{5(15)^2}{2(71)} + \frac{(0.3)(350)(15)}{71} = \$248$$

However, the total cost of the strategy, excluding the CD term, that was actually adopted is

$$TC(Q^*, S^*_{\max}) = \$179$$

Therefore, with a 100% error in estimating the order cost, the actual total cost is higher by

$$\frac{248 - 179}{179} = 38.5\%$$

Through analysis of this type we may conclude whether or not the total cost is sensitive to estimation of the relevant costs. In this analysis we have excluded the CD term because it is a constant, not affected by the choice of Q. In cases where C is a function of Q (e.g., quantity discounts), the term should be included in the sensitivity analysis. The purpose of the sensitivity analysis is to show the sensitivity of *those costs affected by changes in the value of the policy variables.*

3.6.3. QUANTITY DISCOUNTS

So far we have considered a fixed price per unit regardless of the size of the order quantity. Frequently, however, price breaks (discounts) are offered for larger orders. Consider the following price schedule:

Order Size	Unit Variable Cost
$0 < Q < q_1$	C_1
$q_1 \leq Q < q_2$	C_2
$q_2 \leq Q < q_3$	C_3
$q_3 \leq Q$	C_4

The carrying cost and item cost are functions of the price (or cost) of the ordered quantity. The total annual cost expression must include the effect of this discount. Suppose that we are interested in determining the optimal order quantity Q^*, subject to a price break, for a simple inventory model with infinite production rate where shortages are not allowed. The total annual cost equation that describes this situation is

$$\text{TC}(Q) = \frac{AD}{Q} + C_j D + iC_j \frac{Q}{2} \tag{3.23}$$

The decision process for estimating the optimal order quantity in the presence of this price break is quite involved, and it is helpful to develop the analysis as a sequence of decisions. They are:

—*Step 1:* Determine the economic order quantity Q^* based on least price C_4. If $Q^* < q_3$, then Q^* within the region of C_4 equals q_3. The total cost TC (at C_4) is calculated by Eq. (3.23). Go to Step 2. If $Q^* \geq q_3$, then Q^* is optimal within the region C_4. The total cost TC (at C_4) is calculated by Eq. (3.23). Go to Step 2.

—*Step 2:* Determine the economic order quantity Q^*, based on C_3. If

$Q^* < q_2$, then q_2 is the optimal order quantity at C_3. The total cost is evaluated as in Step 1: TC(C_3). Go to Step 3. If $q_2 \le Q^* \le q_3$, then Q^* is optimal within the region C_3. Evaluate the total cost, TC(C_3) and go to Step 3. If $q_3 \le Q^*$, then q_3 is optimal within the region C_3. Evaluate TC(C_3) and go to Step 3.

—Step 3: Repeat Step 2 for the base price C_2. Evaluate TC(C_2).
—Step 4: Repeat Step 2 for the base price C_1. Find TC(C_1).
—Step 5: Compare TC(C_i) for $i = 1, 2, 3, 4$.
—Step 6: Order the quantity Q corresponding to the minimum of TC(C_i).

EXAMPLE 3-3

An operating telephone company purchases large quantities of semiconductors to be used in manufacturing electronic switching systems. Shortages are not allowed. The demand rate is 250,000 units per year, and the order cost is $100 per order. The annual inventory carrying rate is 0.24 and is based on the value of the average inventory. The supplier's price schedule is as follows:

Order Size	Variable Cost Per Unit
$0 < Q < 5000$	$12
$5000 \le Q < 20{,}000$	11
$20{,}000 \le Q < 40{,}000$	10
$40{,}000 \le Q$	9

Determine the optimal order quantity.

SOLUTION

We assume an infinite replenishment rate. Therefore, the total annual cost of the inventory is

$$\text{TC}(Q_j \text{ at } C_j) = \frac{AD}{Q_j} + C_j D + iC_j \frac{Q_j}{2}$$

$$= \frac{100(250{,}000)}{Q_j} + 250{,}000C_j + (0.24C_j)\frac{Q_j}{2} \qquad j = 1, 2, 3, 4$$

(3.24)

The optimal quantity Q_j^* is obtained by equating $dT(Q_j \text{ at } C_j)/dQ_j = 0$.

$$Q_j^* = 10^3 \sqrt{\frac{25}{0.12C_j}}$$

(3.25)

Following Steps 1 through 6, as presented earlier, we obtain:

1. For $j = 4$ and $C_4 = \$9$:

$$Q_4^* = 4811 < 40{,}000$$

Therefore, Q^* is set to equal 40,000.

$$TC(Q^* \text{ at } C_4) = \$2.293.285$$

2. For $j = 3$ and $C_3 = \$10$:

$$Q_3^* = 4564 < 20,000$$

Therefore, Q^* is set at 20,000 units.

$$TC(Q^* \text{ at } C_3) = \$2,525,250$$

3. For $j = 2$ and $C_2 = \$11$:

$$Q_2^* = 4352 < 5000$$

Therefore, Q^* is set at 5000 units.

$$TC(Q^* \text{ at } C_2) = \$2,761,600$$

4. For $j = 1$ and $C_1 = \$12$:

$$Q^* = 4166 < 5000 \text{ units}, \qquad \text{which is in the given range.}$$

Therefore, $Q^* = 4166$ units and

$$TC(Q^* \text{ at } C_1) = 3,012,000$$

5. We compare the results as follows:

Q^*	C_j	$TC(Q^* \text{ at } C_j)$
40,000	9	$\$2,293,285 \leftarrow$ minimum
20,000	10	2,525,250
5,000	11	2,761,600
4,166	12	3,012,000

6. Therefore, the optimal order quantity is 40,000 units at a total cost of $2,337,025.

3.6.4. MULTI-ITEM INVENTORY SYSTEMS WITH CONSTRAINTS

In practice most inventory systems accommodate more than one type of item. Under these situations, the analysis of the inventory problem is approached by treating each type of item independently, and the economic order quantities, optimal shortage, and reorder points can be estimated using the modeling techniques presented earlier. The problem becomes complicated when budgetary, space, and availability-of-item constraints are imposed on the inventory system. These constraints will certainly affect the size of the order quantities for each type of item.

The total annual cost of a multi-item inventory system is estimated as the sum of the total annual cost of each item independently. We consider an n-item continuous review model where shortages are not allowed. The total annual cost of the inventory system for all items is

$$\text{TC}(Q_1, Q_2, \ldots, Q_n) = \sum_{j=1}^{n} \left(C_j D_j + \frac{A_j D_j}{Q_i} + i_j C_j \frac{Q_j}{2} \right) \tag{3.26}$$

The model above may be subject to constraints: for example, a budgetary constraint which requires that at any point in time, no more than a certain amount of money B can be invested in inventory. This can be expressed as

$$\sum_{j=1}^{n} C_j Q_j \leq B \tag{3.27}$$

Other constraints can also be imposed on the inventory system. This problem is formulated as:

$$\text{Minimize TC} = \sum_{j=1}^{n} \text{TC}(Q_j) = \sum_{j=1}^{n} \left(\frac{A_j D_j}{Q_j} + i_j \frac{C_j Q_j}{2} \right) \tag{3.28}$$

Subject to

$$\sum_{j=1}^{n} C_j Q_j \leq B \tag{3.29}$$

$$Q_j \geq 0$$

where TC = total annual inventory cost for all items
  B = maximum investment allowed in inventory

This nonlinear programming problem has been dealt with by using two solution procedures: the *Lagrangian procedure* and the fixed-cycle (*equal-order-interval*) method. The Lagrangian procedure assumes implicitly that orders are received simultaneously, and does not consider phasing orders for the various items. The fixed-cycle method allows the phasing of orders for different items but adds the constraint that all items have the same cycle length. The two procedures [13] are presented below.

The Lagrangian Method: The problem (3.28) is solved by initially ignoring the constraint (3.29) and the optimal ordered quantities are obtained by (assuming that $i_j = i$)

$$Q_j^* = \sqrt{\frac{2A_j D_j}{iC_j}} \qquad j = 1, 2, \ldots, n \tag{3.30}$$

A check on feasibility is made by substituting Q_j^* into constraint (3.29). If the constraint is satisfied, the optimal order quantities are those obtained by (3.30); otherwise, the Lagrangian method is used. This is accomplished by developing a Lagrangian expression (LE), which is given by

$$\text{LE}(Q_j, \lambda) = \sum_{j=1}^{n} \left(\frac{A_j D_j}{Q_j} + \frac{i}{2} C_j Q_j \right) + \lambda \left(\sum_{j=1}^{n} C_j Q_j - B \right) \tag{3.31}$$

Taking the derivatives of Eq. (3.31) with respect to Q_j, λ, and solving the resulting equations after equating them to zero, we obtain

$$Q_{Lj}^* = \sqrt{\frac{2A_j D_j}{C_j(i + 2\lambda^*)}} \tag{3.32}$$

where Q_{Lj}^* is the optimal order quantity obtained by using the Lagrangian method. The value of λ^* is given as

$$\lambda^* = \frac{1}{2} \left(\frac{1}{B} \sum \sqrt{2A_j D_j C_j} \right)^2 - \frac{i}{2} \tag{3.33}$$

Substituting into Eq. (3.32) and rearranging gives

$$Q_{Lj}^* = \frac{B}{\sum_{j=1}^{n} C_j Q_j} Q_j^* = \frac{B}{E} Q_j^* \tag{3.34}$$

where Q_j^*s are those obtained by Eq. (3.30) and

$$E = \sum_{j=1}^{n} C_j Q_j^* \tag{3.35}$$

Equation (3.34) indicates that for this type of model and budgetary constraint, there is no need to solve the Lagrangian problem. Instead, each value of Q_j^* obtained by using Eq. (3.30) should be multiplied by the factor B/E [12]. This implies that the ordered quantities should be reduced by the same factor if there is an excess demand for money over the budget.

It should be noted that the results above are obtained for any linear inequality constraints of the type

$$\sum_{j=1}^{n} a_j Q_j \leq B$$

where $a_j = \gamma C_j$ and γ is a nonnegative constant.

Fixed-Cycle (Equal-Order-Interval) Method: This method assumes a fixed-cycle time for all items (i.e., the orders for n items are phased within the cycle). Let T be the cycle length; then $Q_j = TD_j$. Equation (3.28) is reformulated to be

$$\text{Minimize TC}(T) = \sum_{j=1}^{n} \frac{A_j}{T} + \frac{i}{2} \left(\sum_{j=1}^{n} C_j D_j \right) T \qquad (3.36)$$

The optimal solution of (3.36) without budgetary constraints but imposing fixed cycle time constraint is

$$T^* = \sqrt{\left(2 \sum_{j=1}^{n} A_j \right) \Big/ \left(i \sum_{j=1}^{n} C_j D_j \right)} \qquad (3.37)$$

To determine the optimal solution of (3.36) taking the budgetary constraint (3.29) into account, we let T_0 be the maximum length of the cycle time allowed by the budgetary constraint. Since we assume repetitive cycles of length T_0, it suffices to guarantee that within one (and hence each) cycle the budgetary constraint is satisfied. Let

t_j = time interval from the replenishment of product $j - 1$ to the replenishment of product j

t_1 = time interval from the replenishment of product n to the replenishment of product 1

Then the problem of finding the maximum cycle length T_0 can be formulated as a simple linear programming model [13]:

$$\text{Maximize } T_0 = \sum_{j=1}^{n} t_j \qquad (3.38)$$

subject to

$$\sum_{i=j+1}^{n} \alpha_i \sum_{k=j+1}^{i} t_k + \sum_{i=1}^{j} \alpha_i \left(\sum_{k=1}^{i} t_k + \sum_{k=j+1}^{n} t_k \right) \leq B \qquad \forall \quad j = 1, \ldots, n \qquad (3.39)$$

where $\alpha_i = C_i D_i$. The n budgetary constraints correspond to each of the ordering points which are peaks in terms of budgetary usage.

Due to the characteristics of this problem, at optimality all inequalities hold as equalities (i.e., the budget is exhausted at each peak) and the optimal solution is

$$T_0 = 2B \frac{\sum\limits_{j=1}^{n} \alpha_j}{\sum\limits_{j=1}^{n} \alpha_j^2 + \left(\sum\limits_{j=1}^{n} \alpha_j\right)^2} \tag{3.40}$$

and the optimal phasing of orders for this maximal cycle length, T_0, is

$$t_i = \frac{T_0(\alpha_j)}{\sum\limits_{k=1}^{n} \alpha_k} \qquad j = 1, 2, 3, \ldots, n \tag{3.41}$$

As $TC(T)$ is a convex function, the optimal cycle time T^{**} for the budgetary problem is

$$T^{**} = \min(T^*, T_0) \tag{3.42}$$

EXAMPLE 3-4

Aladdin's Company produces electronic lamps (rather than the usual magic lamps). Three types of components are required for the production of the lamps. The management has an upper limit on the investment of \$16,000. The inventory carrying cost rate for each item is 0.18 and shortages are not allowed. Data for each item are given in Table 3-1. Determine the optimal lot size for each item and the optimal cycle time.

Table 3-1.

	Item 1	Item 2	Item 3
Demand D_j	1500	1500	2500
Item cost C_j	\$60	\$30	\$80
Setup cost A_j	\$60	\$60	\$60

SOLUTION

As mentioned earlier, in solving multi-item inventory problems subject to constraints, one should start by solving the problem ignoring the constraints. After finding the optimal lot sizes, one should then substitute these values into the constraints. If satisfied, no further work is needed. However, if the constraints are not satisfied, the Lagrangian method or the fixed-cycle method should be utilized. The optimal lot sizes are found by using Eq. (3.30):

$$Q_1^* = \sqrt{\frac{2A_1 D_1}{iC_1}} = \sqrt{\frac{2(60)(1500)}{(0.18)(60)}} = 129 \text{ units}$$

$$Q_2^* = \sqrt{\frac{2(60)(1500)}{(0.18)(30)}} = 183 \text{ units}$$

$$Q_3^* = \sqrt{\frac{2(60)(2500)}{(0.18)(80)}} = 144 \text{ units}$$

The inventory investment corresponding to these lot sizes is

$$60(129) + 30(183) + 80(144) = \$24,750 > \$16,000$$

Since the investment constraint is not satisfied, we use Eq. (3.34) to find the optimal order quantities that satisfy this constraint.

$$Q_{Lj}^* = \frac{B}{E} Q_j^*$$

where

$$E = \sum_{j=1}^{n} C_j Q_j^* = \$24,750$$

$$Q_{L_1}^* = \frac{16,000}{24,750} Q_1^* = \frac{16,000}{24,750}(129) = 83 \text{ units}$$

$$Q_{L_2}^* = 118 \text{ units}$$

$$Q_{L_3}^* = 93 \text{ units}$$

and T^* without budgetary constraint but imposing fixed cycle time constraint is obtained from Eq. (3.37):

$$T^* = \sqrt{\frac{2\sum_{j=1}^{n} A_j}{i\sum_{j=1}^{n} C_j D_j}}$$

$$= \sqrt{\frac{2(180)}{(0.18)(335,000)}} = 0.0770 \text{ year}$$

or

$$T^* = 28 \text{ days}$$

T_0 under budgetary and fixed cycle time constraints is

$$T_0 = \frac{2(16,000)(335000)}{(1.6235)(10^{11})} = 0.0660 \text{ year}$$

or

$$T_0 = 24 \text{ days}$$

and

$$T^{**} = \min \{24, 28\} = 24 \text{ days}$$

The optimal order quantities are $Q_1 = 99$, $Q_2 = 99$, and $Q_3 = 165$. Therefore, under the budgetary constraint and the fixed cycle time policy one should order quantities 99, 99, and 165 units for items 1, 2, and 3 respectively at a fixed cycle time of 24 days. From Eq. (3.41), the optimal phasing of orders for this maximal cycle length is $t_1 = 6.4$, $t_2 = 3.2$, and $t_3 = 14.3$ days.

EXAMPLE 3-5

Find the optimal inventory policy for the following three types of products. The pertinent data are shown in Table 3-2. The inventory carrying cost is computed using $i = 0.18$, and shortages are not allowed. The maximum investment at any one point in the inventory is $10,000. Additionally, the warehouse has only 5000 square feet that can be allocated to these products.

TABLE 3-2.

	Product		
	1	*2*	*3*
Demand per year (units)	2000	2200	3800
Cost per item	$8	6	10
Order cost	$70	$70	70
Floor space per item (ft²)	1.0	1.5	3.0

SOLUTION

Following the steps in Example 3-4, we find the optimal lot sizes ignoring the constraints.

$$Q_1^* = \sqrt{\frac{2(70)(2000)}{(0.18)(8)}} = 441 \text{ units}$$

$$Q_2^* = \sqrt{\frac{2(70)(2200)}{(0.18)(6)}} = 534 \text{ units}$$

$$Q_3^* = \sqrt{\frac{2(70)(3800)}{(0.18)(10)}} = 544 \text{ units}$$

If these lot sizes were to be used, the maximum investment in inventory would be

$$8(441) + 6(543) + 10(544) = \$12,226 > \$10,000$$

Therefore, the Lagrangian method should be used. These lot sizes, however, do not violate the floor-space constraint:

$$1(441) + 1.5(534) + 3(544) = 2874 < 5000$$

Therefore, the only effective constraint is the budgetary constraint, which should not exceed $10,000. We now use Eq. (3.34) to find the optimal order quantities that satisfy this constraint.

$$Q_{L_1}^* = \frac{B}{E} Q_1^* = \frac{10,000}{12,226} (441) = 361 \text{ units}$$

$$Q_{L_2}^* = \frac{10,000}{12,226} (534) = 437 \text{ units}$$

$$Q_{L_3}^* = \frac{10,000}{12,226} (544) = 445 \text{ units}$$

EXAMPLE 3-6

Solve Example 3-5 if the maximum allowable space for these products is 2000 ft^2 instead of 5000 ft^2.

SOLUTION

Since the floor-space constraint is the tighter constraint (than the total investment constraint), we estimate the optimal order quantities based on the floor space only.

$$Q_{L_1}^* = \frac{B}{E} Q_1^*$$

where $B = 2000$ and $E = 2874$.

$$Q_{L_1}^* = \frac{2000}{2874} (441) = 307$$

$$Q_{L_2}^* = 372$$

$$Q_{L_3}^* = 378$$

The total floor space is

$$307 + (1.5)(372) + 3(378) = 1999 < 2000$$

and total investment is $8468 < $10,000.

3.7. Probabilistic Inventory Models

In the inventory models developed so far, it is assumed that all parameters of the system are known with absolute certainty. In actual situations one seldom knows exactly what demands will occur over a given period of time. The lead times are also often stochastic. To describe inventory systems subject to uncertainty in the parameters, it is necessary to develop probability distributions for these parameters.

This section presents single-item inventory models in which demands and lead times are stochastic. Following the same objective of deterministic inventory models, the main criterion that will be used is the minimization of the expected inventory costs [carrying cost, order cost (setup cost), and shortage cost].

3.7.1. A CONTINUOUS-REVIEW MODEL

The continuous-review policy has been described at the beginning of this chapter. The inventory level is reviewed continuously and orders of size Q are placed every time the inventory level reaches a reorder level r. The objective is to find the optimum values of Q and r that minimize the total expected inventory cost per unit time. This model is illustrated in Fig. 3-7. Note that the length of the cycles are different.

The following notations are used to describe the model:

D = average demand rate, units/year
h = holding cost per unit per year (iC)
W = shortage cost per unit
A = setup cost per order
x = mean demand over the lead time
$g(x, t)$ = conditional probability density function (p.d.f.) of demand x during lead time t, $x > 0$
$l(t)$ = p.d.f. of lead time t, $t > 0$

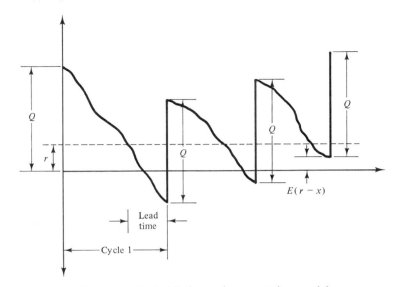

Figure 3-7. Probablistic continuous-review model.

$f(x)$ = p.d.f. of demand x during lead time
Q = amount ordered per cycle
r = reorder inventory level
$S(x)$ = shortage quantity per cycle
$\bar{S}(x)$ = expected shortage per cycle
N = number of orders per year

It should be noted that the absolute p.d.f. of demand x during lead time is

$$f(x) = \int_0^\infty g(x, t)l(t) \, dt$$

The total annual cost of the inventory TC(Q, r) includes the average setup cost, the expected carrying cost, and the expected shortage cost. In the model development that follows, we assume that shortages are backordered.

We must express the total annual cost equation in terms of the decision variables (i.e., in terms of Q and r). The average setup cost per year is AD/Q. The expected holding cost is calculated based on the average inventory level per cycle, which is found as follows.

The expected inventory level at the beginning of a cycle is

$$Q + E[r - x]$$

where $E[r - x]$ is the expected inventory level at the end of the cycle. Consequently, the average inventory per cycle is given by

$$I = \frac{(Q + E[r - x]) + E[r - x]}{2} = \frac{Q}{2} + E[r - x] \qquad (3.43)$$

Using $f(x)$ as defined earlier,

$$E[r - x] = \int_0^\infty [r - x]f(x) \, dx = r - E[x]$$

The average inventory on hand is

$$\frac{Q}{2} + r - E[x] \qquad (3.44)$$

The shortage quantity

$$S(x) = \begin{cases} 0 & x \le r \\ x - r & x > r \end{cases}$$

and the expected shortage quantity per cycle $\bar{S}(x)$ is

$$\bar{S}(x) = \int_0^\infty S(x)f(x)\,dx = \int_r^\infty (x-r)f(x)\,dx \tag{3.45}$$

$$\text{expected shortage per year} = \bar{S}(x)(N) = \frac{\bar{S}(x)D}{Q} \tag{3.46}$$

Multiplying Eqs. (3.44) and (3.46) by the corresponding costs and adding the annual setup cost, we get

$$TC(Q, r) = \frac{AD}{Q} + h\left(\frac{Q}{2} + r - E[x]\right) + \frac{WD}{Q}\,\bar{S}(x) \tag{3.47}$$

The optimal values of Q and r are obtained by solving Eqs. (3.48) and (3.49).

$$\frac{\partial TC(Q, r)}{\partial Q} = \frac{-AD}{Q^2} + \frac{h}{2} - \frac{WD\bar{S}(x)}{Q^2} = 0 \tag{3.48}$$

$$\frac{\partial TC(Q, r)}{\partial r} = h - \frac{WD}{Q}\int_r^\infty f(x)\,dx = 0 \tag{3.49}$$

From Eq. (3.48)

$$Q^* = \sqrt{\frac{2D(A + W\,\bar{S}(x))}{h}} \tag{3.50}$$

and from Eq. (3.49)

$$\int_{r*}^\infty f(x)\,dx = \frac{hQ^*}{WD} \tag{3.51}$$

There are no explicit expressions for Q^* and r^*. Therefore, the solutions for Q^* and r^* can be obtained numerically. The following iterative approach can be used.

Iterative Steps for Finding Optimal r^* and Q^*

1. Let $\bar{S}(x) = 0$ and compute $Q^* = Q_1 = \sqrt{2AD/h}$, where the subscript on Q indicates the iteration number.
2. Use Eq. (3.51) to find the value of r_i corresponding to Q_i.
3. Use r_i in Eq. (3.45) to obtain $\bar{S}(x)_i$, which is used in Eq. (3.50) to obtain a new value Q_i.
4. Compute r_i from Eq. (3.51) by using the value of Q_i obtained in step 3.

5. Repeat steps 3 and 4 until two successive values of r and Q are approximately equal.
6. The last values for Q and r computed in step 5 will yield the optimal values of Q^* and r^*.

EXAMPLE 3-7

A company purchases air filters that are used at the rate of 800 per year. The cost of each filter is $25 and the cost of placing an order is $10. The inventory carrying cost is $2 per unit per year. The shortage cost is $5 per unit. Assume that the demand during lead time follows a uniform distribution over the range 0 to 200. Find Q^* and r^*.

SOLUTION

From Eq. (3.45)

$$\bar{S}(x) = \int_r^\infty (x - r)f(x)\,dx = \int_r^{200} (x - r)\frac{1}{200}\,dx$$

$$= \frac{r^2}{400} - r + 100 \tag{3.52}$$

Using Eq. (3.50), we get

$$Q^* = \sqrt{\frac{2D(A + W\,\bar{S}(x))}{h}} = \sqrt{\frac{2(800)(10 + 5\bar{S}(x))}{2}}$$

$$= \sqrt{8000 + 4000\bar{S}(x)} \tag{3.53}$$

From Eq. (3.51)

$$\int_{r*}^{200} \frac{1}{200}\,dx = \frac{2Q^*}{5(800)} = \frac{Q^*}{2000}$$

which yields

$$r^* = 200 - \frac{Q^*}{10} \tag{3.54}$$

Equation (3.54) is used to compute r_i for a given value of Q_i. This value of r_i is then used in Eq. (3.52) to compute $\bar{S}_i(x)$. The value of Q_{i+1} corresponding to r_i and $\bar{S}_i(x)$ is obtained from Eq. (3.53).

Iteration 1:
An initial value of Q is obtained as

$$Q_1 = \sqrt{\frac{2AD}{h}} = \sqrt{\frac{2(800)(10)}{2}} = 89 \text{ units}$$

$$r_1 = 200 - \frac{89}{10} = 191.1 \text{ units}$$

Iteration 2:

$$\bar{S}_1(x) = \frac{r_1^2}{400} - r_1 + 100 = 0.1980$$

$$Q_2 = \sqrt{8000 + 4000(0.198)} = 93.76$$

$$r_2 = 200 - \frac{93.76}{10} = 190.624$$

Iteration 3:

$$\bar{S}_2(x) = \frac{r_2^2}{400} - r_2 + 100 = 0.2197$$

$$Q_3 = \sqrt{8000 + 4000(0.2197)} = 94.228$$

$$r_3 = 200 - \frac{94.228}{10} = 190.577$$

Iteration 4:

$$\bar{S}_3(x) = \frac{r_3^2}{400} - r_3 + 100 = 0.22198$$

$$Q_4 = \sqrt{8000 + 4000(0.22198)} = 94.275$$

$$r_4 = 200 - \frac{94.275}{10} = 190.5725$$

Since r_3 and r_4 are approximately equal,

$$r^* = 190 \qquad \text{and} \qquad Q^* = 94$$

3.7.2. PERIODIC-REVIEW MODELS

In Section 3.4.1 the periodic-review policy is described in which the inventory levels are observed at equal intervals of time T and orders are placed when the inventory level becomes less than or equal to a predetermined reorder level. A widely used version of this model is the order up to I_{max} policy, in which the reorder point is I_{max}. The objective of this section is to determine the optimal values of T^* and I_{max}^* which minimize the total annual inventory cost when using the order up to I_{max} policy. A typical periodic-review model is shown in Fig. 3-8.

In the development of the stochastic continuous-review model we assumed that shortages were backordered. For contrast, in this section we assume that shortages represent lost sales. It is also assumed that an order placed at time i will be received l time units later. In addition to the notations given in Section 3.7.1, we define

V = cost of making a review
l = average lead time $[\int l(t)\, dt]$

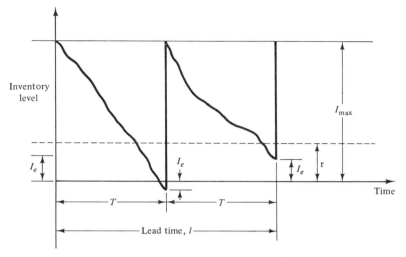

Figure 3-8. Periodic-review model.

I_e = on-hand inventory at the end of a typical period before the next order arrives

I_s = on-hand inventory at the start of a typical period after the order has arrived

Following the same procedures as those of Section 3.7.1, we obtain the following inventory costs.

1. *Review cost and order cost:* Since a review takes place every T units of time, there will be $(1/T)$ reviews per year. If we treat demand as continuous, it is reasonable to assume that an order will be placed at each review and the total review and order cost is $(V + A)/T$.
2. *Carrying cost:* The expected inventory level at the end of the cycle is

$$E[I_e] = \int_0^\infty (I_{\max} - x)g(x, l + T)\, dx$$

$$+ \int_{I_{\max}}^\infty (x - I_{\max})g(x, l + T)\, dx$$

$$= I_{\max} \int_0^\infty g(x, l + T)\, dx - \int_0^\infty xg(x, l + T)\, dt \quad (3.55)$$

$$+ \int_{I_{\max}}^\infty (x - I_{\max})g(x, l + T)\, dx$$

$$= I_{\max} - E[x, l + T] + \bar{S}(I_{\max}, T)$$

$$E[I_e] = I_{\max} - D(l + T) + \bar{S}(I_{\max}, T) \quad (3.56)$$

where $\bar{S}(I_{max}, T)$ is the lost demand (sales) per period. But the expected on-hand inventory level at the start of a cycle is

$$E[I_s] = E[I_e] + DT \qquad (3.57)$$

The average inventory level I can be obtained as

$$I = E[I_e] + \tfrac{1}{2}(E[I_s] - E[I_e])$$
$$I = I_{max} - Dl - \tfrac{1}{2}DT + \bar{S}(I_{max}, T) \qquad (3.58)$$

The average carrying cost is

$$hI = h[I_{max} - Dl - \tfrac{1}{2}DT + \bar{S}(I_{max}, T)] \qquad (3.59)$$

3. *Cost of lost sales:* Again, shortage occurs only when the demand exceeds the target inventory level; therefore,

$$\bar{S}(I_{max}, T) = \int_{I_{max}}^{\infty} (x - I_{max})g(x, l + T)\, dx \qquad (3.60)$$

The average annual cost of shortages incurred per year is

$$\frac{W\,\bar{S}(I_{max}, T)}{T} \qquad (3.61)$$

The total annual cost, $\text{TC}(I_{max}, T)$, is obtained by adding Eqs. (3.59) and (3.61) to the total review and order cost:

$$\text{TC}(I_{max}, T) = \frac{V + A}{T} + h[I_{max} - Dl - \tfrac{1}{2}DT + \bar{S}(I_{max}, T)]$$
$$+ \frac{W\,\bar{S}(I_{max}, T)}{T} \qquad (3.62)$$

The optimal values of I_{max} for a given T can be obtained by setting $\partial\text{TC}(I_{max}, T)/\partial I_{max}$ and $\partial\text{TC}(I_{max}, T)/\partial T$ to zero and solving simultaneously. The latter derivate cannot be explicitly obtained. Therefore,

$$\frac{\partial\text{TC}(I_{max}, T)}{\partial I_{max}} = h + \left(h + \frac{W}{T}\right)\frac{\partial}{\partial I_{max}}\bar{S}(I_{max}, T) = 0 \qquad (3.63)$$

which yields

$$\int_0^{I_{max}^*} g(x, l + T)\, dx = \frac{W}{W + hT} \qquad (3.64)$$

An iterative approach to the solution of this problem is to assume values for T and solve for the corresponding optimal value of I_{max}. A procedure that converges on I^*_{max} and T^* is given in the example that follows.

EXAMPLE 3-8

An assembly plant of a small automobile manufacturer is using a periodic-review inventory model (order up to I_{max}) to place orders for tires needed for the car assembly. The demand rate is 12,000 tires per year. The average lead time l is 6 months. The cost of making a review is $20 and the cost of placing an order is $30. The cost of each tire is $40 and the inventory carrying cost rate is 0.20 per year. The shortage cost per unit is $25. The demand during $l + T$ is represented by a normal distribution with a mean of $12,000(l + T)$ and variance of $8000(l + T)$. Determine the optimal values of I^*_{max} and T^*.

SOLUTION

This problem could be solved by using Eq. (3.64) to calculate I_{max} for a given T_i. The total inventory cost corresponding to I_{max} and T_i is estimated by Eq. (3.62). The value of T_i may be incremented by $\pm \Delta T$ and I_{max} and total inventory cost are estimated. Search out I^*_{max} and T^* by moving in decreasing cost direction.

$$D = 12,000 \text{ units/year}$$

$$C = \$40 \qquad\qquad i = 0.20 \qquad W = \$25$$

$$A = \$30 \qquad\qquad V = \$20 \qquad l = \tfrac{1}{2} \text{ year}$$

Let $T_1 = 3$ months and by using Eq. (3.64), the value of I^*_{max} is obtained as follows:

$$\int_0^{I^*_{max}} g(x, l + T_1)\, dx = \frac{W}{W + hT_1} = \frac{25}{25 + (40)(0.2)(\tfrac{1}{4})} = 0.9259$$

Since $T_1 = \tfrac{1}{4}$ year, the expected demand in $l + T_1$ is $12,000(\tfrac{1}{2} + \tfrac{1}{4}) = 9000$ units, and the variance of the demand during this time is $8000(\tfrac{3}{4}) = 6000$. Therefore,

$$\frac{I^*_{max} - \mu}{\sigma} = Z(0.9259)$$

where μ and σ are the mean and standard deviations of the demand during $l + T_1$, respectively. The number of standard deviations corresponding to $Z(0.9259)$ is obtained from the tables of standard normal distribution.

$$\frac{I^*_{max} - 9000}{\sqrt{6000}} = 1.45$$

$$I^*_{max} = 9,112.3 \text{ tires}$$

From Eq. (3.60) the average shortage (lost sales) is

$$\bar{S}(I_{max}, T) = \int_{I_{max}}^{\infty} (x - I_{max})g(x, l + T)\, dx$$

$$= \sigma\phi(z) - [(I_{max} - \mu)(1 - \Phi(z))] \tag{3.65}$$

where $\phi(z)$ = ordinate under the normal distribution density function

$$= (1/\sqrt{2\pi})e^{-z^2/2}$$

$\Phi(z)$ = cumulative area under the normal distribution

A proof of Eq. (3.65) is given in Appendix A.

$$\bar{S}(9112.3, \tfrac{1}{4}) = 77.45\phi(1.45) - (9112.3 - 9000)[1 - \Phi(1.45)]$$

$$= (77.45)(0.1394) - (112.3)(0.0741) = 2.476$$

Using Eq. (3.62), the total inventory cost is

$$TC(9112.3, \tfrac{1}{4}) = \frac{50}{\tfrac{1}{4}} + 8[9112.3 - 12,000(\tfrac{1}{2}) - \tfrac{1}{2}(12,000)(\tfrac{1}{4})$$

$$+ 2.476] + \frac{25}{0.25}(2.47) = \$13,365$$

Increment T_1 by 0.5 month to be 2.5 months and repeat the steps above. For $T_2 = 2.5$ months (0.2083 year),

$$\mu = 12,000(\tfrac{1}{2} + 0.2084) = 8500$$

$$\sigma^2 = 8000(\tfrac{1}{2} + 0.2084) = 5666$$

$$I_{max}^* = \mu + \sigma Z(0.9375) = 8614.4$$

$$\bar{S}(8614.4, 0.2084) = 75.27(0.1233) - (8614.4 - 8500)(0.0625)$$

$$= 2.13$$

$$TC(8614.4, 0.2084) = \$11,390.40$$

The steps above are repeated for other values of T. A summary of the results is shown in Table 3-3. The optimal T^* and I_{max}^* are 15 days and 6645 tires at a cost of \$4590.

Table 3-3.

$T_1(days)$	I_{max}^*	$TC(I_{max}^*, T_1)$
90	9112	\$13,365
75	8614	11,390
60	8119	9,498
45	7625	7,625
15	6645	4,590*
5	6334	6,041

3.7.3. SINGLE-PERIOD MODELS

These models are special cases of the periodic-review models presented in Section 3.7.2. The single-period models exist when an item is ordered only once to satisfy the demand of a given period of time (i.e., there is no replenishment of the item during the period and backlogging is not allowed). Typical examples of items that could be classified under single-period inventory models are Christmas trees, garden tools, snow tires, and winter overcoats. Analyses of single-period inventory models for instantaneous and uniform demand are presented.

Instantaneous demand. In this inventory model, it is assumed that the demand rate is infinite and occurs at the beginning of the period. It is assumed that the setup cost is negligible. The following two situations may occur (see Fig. 3-9):

1. The expected demand is less than the ordered quantity; consequently, no shortage will occur.
2. The expected demand is greater than the ordered quantity; consequently, a shortage will occur. Again, we must develop a total inventory cost equation, its components being:
   (a) Holding cost (carrying cost): Let Q be the quantity on hand at the beginning of the period after an order is received and δ is the demand that occurs at the beginning of the period. The inventory level, $\text{IL}(Q)$ is

$$\text{IL}(Q) = \begin{cases} Q - \delta & \text{for } \delta < Q \\ 0 & \text{for } \delta \geq Q \end{cases} \tag{3.66}$$

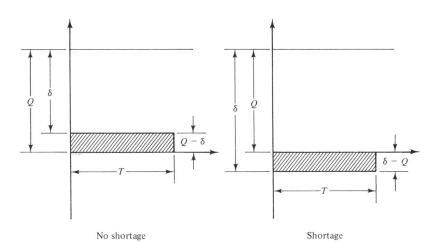

No shortage   Shortage

Figure 3-9. Single-period inventory model with instantaneous demand.

The expected holding cost is

$$E[\text{holding cost}] = h \int_0^Q \text{IL}(Q)\phi(\delta) \, d\delta \qquad (3.67)$$

where $\phi(\delta)$ = p.d.f. of demand
h = holding cost per unit per period

(b) Shortage inventory cost: The amount of shortage of inventory is

$$\text{SI}(Q) = \begin{cases} 0 & \text{for } \delta \leq Q \\ \delta - Q & \text{for } \delta > Q \end{cases} \qquad (3.68)$$

and the expected shortage cost during the period is

$$E[\text{shortage cost}] = W \int_Q^\infty (\delta - Q)\phi(\delta) \, d\delta \qquad (3.69)$$

where W is the shortage cost per unit short.

(c) Total item cost:

$$\text{item cost} = C(Q - x) \qquad (3.70)$$

where C = purchase price per unit
x = amount on hand before an order is received

The expected total inventory cost $E[TC(Q)]$ is thus the sum of Eqs. (3.67), (3.69), and (3.70).

$$E[TC(Q)] = C(Q - x) + h \int_0^Q (Q - \delta)\phi(\delta) \, d\delta + W \int_Q^\infty (\delta - Q)\phi(\delta) \, d\delta \qquad (3.71)$$

The optimum value of Q is obtained by equating $dE[TC(Q)]/dQ$ to zero and solving for Q^*.

$$\frac{dE[TC(Q)]}{dQ} = C + h \int_0^Q \phi(\delta) \, d\delta - W \int_Q^\infty \phi(\delta) \, d\delta = 0 \qquad (3.72)$$

Since $\int_Q^\infty \phi(\delta) \, d\delta = 1 - \int_0^Q \phi(\delta) \, d\delta$, Eq. (3.72) reduces to

$$\int_0^{Q^*} \phi(\delta) \, d\delta = \frac{W - C}{W + h} \qquad (3.73)$$

Q^* exists if $W \geq C$. The value of Q^* is that Q for which $P(\delta \leq Q) = (W - C)/(W + h)$.

EXAMPLE 3-9

Consider the case where a one-period inventory model is applicable. Assume that $h = \$2.00$, $W = \$10.00$, and $C = \$2.50$. The demand density function is given by

$$\phi(\delta) = \begin{cases} \frac{1}{20} & 0 \leq \delta \leq 20 \\ 0 & \delta > 20 \end{cases}$$

Determine the optimal order quantity.

SOLUTION

$$P(\delta \leq Q^*) = \frac{W - C}{W + h} = \frac{(10.0 - 2.5)}{10.0 + 2} = 0.625$$

Using Eq. (3.73), we have

$$P(\delta \leq Q^*) = \int_0^{Q^*} \phi(\delta) \, d\delta = \int_0^{Q^*} \frac{1}{20} \, d\delta = \frac{Q^*}{20}$$

which yields $Q^* = 12.5$ units. If the demand is described by a discrete probability distribution rather that a continuous distribution, Eq. (3.71) becomes

$$E[TC(Q)] = C(Q - x) + h \sum_{\delta=0}^{Q} (Q - \delta)\phi(\delta) + W \sum_{\delta=Q+1}^{\infty} (\delta - Q)\phi(\delta) \qquad (3.74)$$

The conditions for optimality of Eq. (3.74) are

$$E[TC(Q - 1)] \geq E[TC(Q)] \leq E[TC(Q + 1)]$$

It could be shown that the optimal value of Q^* can be found if

$$P(\delta \leq Q^* - 1) \leq \frac{W - C}{W + h} \leq P(\delta \leq Q^*) \qquad (3.75)$$

EXAMPLE 3-10

Determine optimal order quantity Q for the data given in Example 3-9 when the demand distribution function is given by Table 3-4.

Table 3-4.

δ	0	1	2	3	4
$\phi(\delta)$	0.15	0.20	0.30	0.20	0.15

SOLUTION

The critical ratio of Eq. (3.75) is

$$P = \frac{W - C}{W + h} = \frac{10.0 - 2.5}{10.0 + 2.5} = 0.625$$

The cumulative distribution function of the demand is as shown in Table 3-5.

Table 3-5.

δ	0	1	2	3	4
$P(\delta \leq Q)$	0.15	0.35	0.65	0.85	1.0

$$\uparrow$$
$$P = 0.625$$

$$P(\delta \leq 1) = 0.35 < 0.625 < 0.65 = P(\delta \leq 2)$$

or $Q^* = 2$ units.

Uniform demand. This model is similar to the instantaneous demand model presented in Section 3.7.3, with the exception that the demand occurs uniformly over the inventory period. Again, either of the following two situations may occur: (1) the demand is less than the starting inventory level, which will result in excess inventory at the end of the period, or (2) the demand exceeds the starting inventory level, which causes a shortage. These situations are shown in Fig. 3-10. The cost components of the total inventory cost equation are as follows:

1. *Holding (or carrying cost):* The carrying cost is calculated based on the average inventory level IL(Q)

$$IL(Q) = \begin{cases} Q - \dfrac{\delta}{2} & \text{for} \quad \delta \leq Q \\[2mm] \dfrac{Q^2}{2\delta} & \text{for} \quad \delta > Q \end{cases}$$

Therefore, the expected inventory carrying cost is

$$h\left[\int_0^Q \left(Q - \frac{\delta}{2} \right) \phi(\delta) \, d\delta + \int_Q^\infty \frac{Q^2}{2\delta} \phi(\delta) \, d\delta \right] \tag{3.76}$$

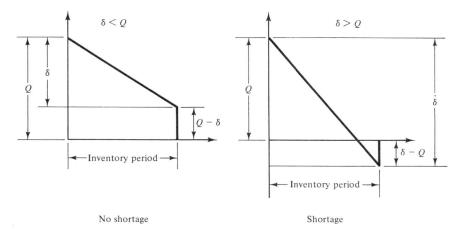

Figure 3-10. Single-period model with uniform demand.

2. *Shortage cost:* The expected shortage of the inventory, SI(Q), is

$$SI(Q) = \begin{cases} \dfrac{(\delta - Q)^2}{2\delta} & \text{for} \quad \delta > Q \\ 0 & \text{for} \quad \delta \le Q \end{cases}$$

and the average shortage cost is given by

$$W \int_Q^\infty \frac{(\delta - Q)^2}{2\delta} \, \phi(\delta) \, d\delta \tag{3.77}$$

3. *Purchase cost:*

$$\text{purchase cost} = C(Q - x) \tag{3.78}$$

The expected total inventory cost is obtained by adding Eqs. (3.76), (3.77), and (3.78).

$$E[TC(Q)] = C(Q - x) + h\left[\int_0^Q \left(Q - \frac{\delta}{2} \right) \phi(\delta) \, d\delta + \int_Q^\infty \frac{Q^2}{2\delta} \phi(\delta) \, d\delta \right]$$

$$+ W \int_Q^\infty \frac{(\delta - Q)^2}{2\delta} \, \phi(\delta) \, d\delta \tag{3.79}$$

The optimal value of Q is obtained by setting

$$\frac{dE[TC(Q)]}{dQ} = 0$$

Thus

$$C + h \left[\int_0^Q \phi(\delta) \, d\delta + \int_Q^\infty \frac{Q}{\delta} \phi(\delta) \, d\delta \right] - W_1 \int_0^\infty \frac{(\delta - Q)}{\delta} \phi(\delta) \, d\delta = 0$$

or

$$\int_0^{Q*} \phi(\delta) \, d\delta + Q* \int_{Q*}^\infty \frac{\phi(\delta)}{\delta} \, d\delta = \frac{W - C}{W + h} \qquad (3.80)$$

EXAMPLE 3-11

Consider Example 3-9 and assume that the demand occurs uniformly over the inventory period. Determine $Q*$.

SOLUTION

Using Eq. (3.80) gives us

$$\int_0^{Q*} \frac{1}{20} \, d\delta + Q* \int_{Q*}^{20} \frac{1}{20\delta} \, d\delta = 0.625$$

$$\tfrac{1}{20}(3.995Q* - Q* \ln Q*) = 0.625$$

$$3.995Q* - Q* \ln Q* - 12.5 = 0$$

By trial and error, $Q*$ is found to be 5.4 units.

3.8. Inventory System Control Practices

The purpose of this section is to present some practical techniques of inventory analysis and control that are often used in conjunction with the models that we have discussed previously.

3.8.1 THE ABC CLASSIFICATION SYSTEM

Vilfredo Pareto, in his book entitled *The Theory of Statistics* (1896), states that "In any series of elements to be controlled, a selected small fraction in terms of number of elements, always accounts for a large fraction, in terms of effect." In the inventory control field, this observation is noticed in the ABC classification of inventory. Under the ABC approach, it is often found that about 20% of the items that are inventoried account for about 80% of the total annual dollar volume (demand × price). The remaining 80% of the items account for only 20% of the total dollar value. In other words, a few products account for the greatest profit potential. These are called *A items*. Some of the items, group B, account for a smaller amount of the dollar volume, and group C accounts for a very small amount of the dollar volume. Figure 3-11 shows

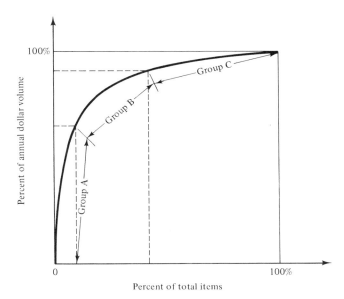

Figure 3-11. ABC analysis.

the ABC classification. The inventory policies for the groups are:

—*Group A:* This group represents about 7 to 10% of the items carried in inventory and 60% of the annual inventory dollar volume. It is recommended that this group of items be given close attention. For example, check the usage and procurement cost monthly and use economic order models for determining lot sizes.

—*Group B:* This group represents about 10 to 20% of the items carried in inventory and about 24% of the annual inventory dollar volume. It is recommended that the monthly usage be checked every 6 months for redetermination of order quantity.

—*Group C:* This group represents about 80% of the items carried in inventory and about 16% of the annual inventory dollar volume. One may check usage annually and simply stock a year's supply. Developing minimum-cost models for all the items in this category will probably be more expensive than the savings that would occur, so a simple ordering rule is preferred in most cases.

PROBLEMS

3-1. The ABC Company estimates the cost of ordering a raw material to be $50 per order. The cost of inspection of the received material is $10 per order. The monthly demand is 3000 units and inventory carrying cost is $4 per unit per month. Find the optimum order quantity for the following independent assumptions:

1. Shortages are not permitted.
2. Shortage cost is $50 per unit short per month.
3. Shortage cost is $10 per unit short independent of the duration.
4. Errors in estimating shortage cost in (2) are +50% to −50%.

Draw a relationship diagram to show the effect of these errors on EOQ.

3-2. A manufacturing facility is planning to increase the storage space by building a new high-bay warehousing system. An analysis is performed for one of the main products of the facility and the following data are obtained:

1. The annual demand for this product is 10,000 units.
2. The cost of providing space is $1 per unit per day.
3. The combined cost of storing or retrieving a unit is $0.50.
4. The inventory carrying cost rate is 20% based on the average inventory level.
5. The cost of shortage is $30 per unit per day.
6. The cost of the product is $40 per unit.
7. The cost of production setup is $200 per order.
8. The production rate is 50,000 units per year.
9. The space required for each unit is 2 ft^3.

How much space should be allocated for this product such that the total inventory cost is minimized?

3-3. Consider an inventory system where discounts are given for large ordered quantities. The price schedule is shown in Table 3-6. The shortage cost is $8 per unit and the order cost is $40 per order. The inventory carrying cost is $2 per unit per year and the annual demand is 5000 units. Determine the optimum order quantity.

3-4. An industrial engineer wishes to determine whether a subassembly of a product is to be made in the plant or to be manufactured by an outside supplier. The data pertinent to the decision-making process are shown in Table 3-7. The carrying cost per unit is $0.50 per day and the shortage cost per unit per day is $0.20. How should the engineer decide on ordering this subassembly

Table 3-6.

$Q < 1000$ units		$C = \$10.0/\text{unit}$
$1000 \leq Q < 2000$ units		$C = \$9.0/\text{unit}$
$2000 \leq Q < 3500$ units		$C = \$8.0/\text{unit}$
$3500 \leq Q$ units		$C = \$7.5/\text{unit}$

Table 3-7.

	In-Plant	Vendor 1	Vendor 2
Production per day (units)	200	400	∞
Lead time (days)	4	6	9
Cost per unit	$50	$55	$54
Order cost	$100	$60	$70

(a) If the annual demand is 30,000 units?

(b) If the annual demand is 20,000 units and the shortage cost per unit is $0.15 independent of the period?

(c) Assuming that the annual demand is 20,000 units and that vendor 1 provides a quantity discount as shown in Table 3-8?

Table 3-8.

$Q < 5000$	$C = \$55/\text{unit}$
$5000 \le Q < 10{,}000$	$C = \$50/\text{unit}$
$10{,}000 \le Q$	$C = \$48/\text{unit}$

3-5. Consider a situation where the procurement rate of a product is infinite and the carrying cost rate is 0.15 per year, based on the maximum inventory level. The shortage cost per unit per year is $1.50 and is based on the average shortage quantity. The procurement cost is $30 per order; shipping and inspection cost is $100 per order. The annual demand is 1000 units and the cost per unit is $5. Determine the optimum order quantity and the optimum reorder level.

3-6. Derive expressions for the optimum order quantity Q^* and reorder r^* for the inventory system described in Section 3.6.1 (assuming that shortage cost per unit is W and independent of the duration).

3-7. Consider an inventory system with a deterministic demand rate D. Shortages are allowed and the carrying cost is based on the maximum inventory level. Find Q^* and S^* (replenishment rate is P).

3-8. The following estimates have been determined for an inventory problem: $D = 2500$ unit/period, $A = \$15$ per period, $i = 18\%$, $C = \$8$ per unit, and the production rate is 3500 units per period. If the shortage cost per unit is negligible and the inventory manager decides on an arbitrary optimum quantity of 500 units, what is the implied shortage per unit per period?

3-9. Using the ABC inventory approach, company XYZ identified three main products in group A. The inventory data for these products are given in Table 3-9.

Table 3-9.

	Product 1	Product 2	Product 3
Demand per year (units)	2000	3000	1500
A	$50	$40	$60
C	$40	$70	$30

Assume that the carrying cost rate is 0.24. What are the optimal quantities to be ordered from these products if the total investment in the inventory does not exceed $8000? What is the optimum cycle time?

3-10. Suppose that an inventory system accommodates the four types of items shown in Table 3-10. The inventory carrying rate is 0.20 per year. The maximum investment is $15,000 and the maximum floor space available is 8000 ft².

(a) What are the optimal ordered quantities of these items? What is the optimal cycle time?

(b) Rework part (a) for a space constraint of 1200 ft².

Table 3-10.

	Item 1	Item 2	Item 3	Item 4
Demand per year (units)	3000	4000	2000	3200
C	$20	$30	$25	$40
A	$100	$110	$105	$80
Floor space per unit (ft²)	3	2	4	2.5

3-11. Solve the inventory problem described by Eqs. (3.28) and (3.29) when shortage costs per unit per period are allowed.

3-12. A manufacturing engineer is to decide on the optimal order quantity for the number of spare dies for three different stamping operations. It is found that the average life of a die is 30,000 stampings and the number of stampings for each die per day is 2000. Other data are given in Table 3-11. The downtime per machine due to lack of dies is $5 per hour. The carrying cost rate is 0.18 and the maximum floor space available is 100 ft². What are the optimal number of spares for each die type?

Table 3-11.

	Die 1	Die 2	Die 3
C	$100/unit	$150/unit	$80/unit
A	$50/unit	$40/unit	$50/unit
Floor space (ft²)	2.0	1.5	1.8

3-13. A machine shop purchases spare parts for several machines at the rate of 1000 per year. The cost of each part is $20, and the cost of placing an order is $20. The inventory carrying cost is $2.5 per unit per year. The shortage cost is $5.5 per unit per year. Assume that the demand (x) during lead time follows an exponential distribution with a p.d.f. of $e^{0.05x}$ over the range 0 to 200. The shortage cost per unit short is $2 based on the average shortage. Find Q^* and r^*.

3-14. Determine the optimal Q^* and r^* for the continuous-review model given in Section 3.7.1. Allow a shortage cost per unit independent of the duration of the shortage.

3-15. Consider a continuous-review inventory model, where the p.d.f. of the demand x during lead time follows a uniform distribution over the range 0 to 100. The annual demand is 600 units and the cost of each unit is \$20. The inventory carrying cost is \$2 per unit per year and the order cost is \$20 per order. Based on the experience, it is found that a reorder level of 80 units could be optimal. What are the optimum order quantity and the implied shortage cost?

3-16. Solve Prob. 3-15 assuming a production rate of 4000 units per year.

3-17. In a typical periodic-review model, it is assumed that the annual demand is 10,000 units. The cost of making a review is \$10 and the cost of placing an order is \$25. The cost per unit is \$20, the carrying cost is \$4 per unit per year, and the shortage cost is \$10 per unit per year. It is found that the demand during $l + T$ is represented by a normal distribution with a mean $10,000(l + T)$ and variance $8000(l + T)$. Determine the optimal values of I^*_{max} and T^* (the lead time is 4 months).

3-18. Develop expressions of an optimal I^*_{max} and T^* for the periodic-review model given in Section 3.7.2 where backorders are not lost (i.e., backlogging is allowed).

3-19. Consider a periodic-review model where the demand rate is 1000 units per year. The carrying cost is \$8 per unit per year. The cost per unit is \$80 and the cost of making a review is \$20. The order cost is \$10 per order. The demand during $l + T$ is uniformly distributed over 200 to 400 units. If the optimal T^* is 45 days and lead time is 4 months, what are the implied shortage cost and the optimum value of I^*_{max} ?

3-20. Consider a single-period inventory model where the inventory level during production increases according to the following expression:

$$I(t) = (P - D)t^2$$

where $I(t) = $ inventory level at time t
$P = $ production rate
$D = $ demand rate

Find the optimal Q^* (shortage is not allowed).

3-21. In a single-period inventory model, the demand density function is given by a normal distribution with mean of 100 units and variance of 60. The carrying cost is \$5 per unit, the shortage cost is \$15, and the cost of the item is \$10. Find Q^*.

3-22. The probability of a demand for a product is given by

x	10	20	30	40	50
$P(x)$	0.20	0.25	0.35	0.10	0.10

The carrying cost is \$5 per unit, the shortage cost is \$15, and the cost of the product is \$8 per unit. Find Q^*.

3-23. In a single-period inventory model, the demand occurs uniformly over the inventory period. The carrying cost is \$2 per unit per year, the shortage cost is \$10, and the item cost is \$8. Assume that the demand follows a uniform distribution over the range 100 to 200. Find Q^*.

3-24. Solve Prob. 3-23 assuming that the demand occurs uniformly over the inventory period.

3-25. Repeat Prob. 3-20 assuming that

$$\phi(\delta) = \begin{cases} e^{\delta} & \delta > 0 \\ 0 & \text{otherwise} \end{cases}$$

and zero initial inventory.

REFERENCES

[1] BUFFA, ELWOOD S., *Production-Inventory Systems: Planning and Control.* Homewood, Ill.: Richard D. Irwin, Inc., (1968) chapters 3 and 4.

[2] ELSAYED, ELSAYED A., AND CHRIS TERESI, "Analysis of Inventory Systems With Deteriorating Items." *International Journal of Production Research*, Vol. 21, No. 4, July/Aug., 1983.

[3] FABRYCKY, WOLTER L., "A Multi-Source Procurement and Inventory Model," *Journal of Industrial Engineering*, XVI, August 1965.

[4] HADLEY, GEORGE, AND THOMAS M. WHITIN, *Analysis of Inventory Systems.* Englewood Cliffs, N.J.: Prentice-Hall, Inc. 1963.

[5] HAUSSMANN, FRED, *Operations Research in Production and Inventory Control.* New York, N.Y.: John Wiley & Sons, 1962.

[6] IMO, I. I. AND D. DAS, "Multi-stage, Multi-facility, Batch Production System With Deterministic Demand Over a Finite Time Horizon," *International Journal of Production Research*, Vol. 21, No. 4, July/Aug., 1983.

[7] JOHNSON, LYNWOOD A., AND DOUGLAS C. MONTGOMERY, *Operations Research in Production Planning, Scheduling and Inventory Control.* New York, N.Y.: John Wiley & Sons, Inc., 1974.

[8] LAU, HON-SHIANG AND AHMED ZAKI, "The Sensitivity of Inventory Decisions to the Shape of the Lead Time Distribution," *IIE Transactions*, Vol. 14, No. 4, Dec. 1982.

[9] MAGEE, JOHN F., AND DAVID M. BOODMAN, *Production Planning and Inventory Control*, 2nd ed., New York, N.Y.: McGraw-Hill Book Co., Inc., 1967.

[10] NADDOR, ELIEZER, *Inventory Systems.* New York, N.Y.: John Wiley & Sons, 1966.

[11] ORAL, MUHITTIN, "Equivalent Formulations of Inventory Control Problems," *AIIE Transactions*, Vol. 13, No. 4, Dec. 1981.

[12] PARSONS, J. A., "Multi-Product Lot Size Determination When Certain Restrictions are Active," *Journal of Industrial Engineering*, 17, 1966.

[13] ROSENBLATT, MEIR J., "Multi-Item Inventory System With Budgetary Constraint: A Comparison Between the Lagrangian and the Fixed Cycle Approach," *International Journal of Production Research*, 19, no. 4, 1981.

[14] SHORE, BARRY, *Operations Management.* New York, N.Y.: McGraw-Hill Book Co., 1973.

[15] STARR, MARTIN K., *Systems Management of Operations*. Englewood Cliffs, N.J.: Prentice-Hall, Inc., chapter 12, 1971.

[16] STARR, MARTIN K., AND DAVID W. MILLER, *Inventory Control: Theory and Practice*. Englewood Cliffs, N.J.: Prentice-Hall, Inc., 1962.

[17] TAHA, HAMDY A., *Operations Research*. New York, N.Y.: The Macmillan Co., 1976.

AGGREGATE PRODUCTION PLANNING

4.1. The Purpose of Aggregate Production Planning

Customer demand enters the production system as units of products. However, production has to be planned as hours of machining and worker-hours of work that must be dedicated to the production of that demand. In this chapter we describe some methods by which this planning is done.

When planning work-force and related activities to service a given demand schedule, it is necessary to balance the cost of building and holding inventory against the cost of adjusting activity levels to fluctuations in demand. Figure 4-1 illustrates a hypothetical cumulative demand pattern and two alternative production strategies. Alternative 1 uses a constant work-force level (i.e., constant production output rate). Since the production output rate is greater than the expected demand rate in the earlier production periods, cumulative production will exceed cumulative demand, resulting in a significant inventory carrying cost. Conversely, significant shortage cost may incur when the cumulative demand exceeds the cumulative production.

Alternative 2 is a strategy to produce to demand such that the inventory carrying costs are minimized. This alternative requires constantly adjusting the work-force levels or paying significant overtime cost during the high demand periods.

4

These are two extreme alternatives: the optimal alternative is the one that minimizes the total cost of the inventory and the cost of adjusting the work-force level. In this chapter we present approaches for computing aggregate plans that could respond to anticipated demand fluctuations while attempting to incur a minimum overall cost of production. The primary output of the aggregate planning process is a *master schedule*, which describes the number of units to be produced during each period and the work-force levels required by period.

The approach for finding the optimal alternative (master schedule and work-force level) is to develop a total cost function which contains the major cost component of the production facility. This cost function is to be minimized while subject to constraints. The linearity or nonlinearity of the cost function and constraints determine the solution approach to the problem. For example, linear programming can be used for solving aggregate production planning problems with a linear function and constraints. Other approaches are used when the function or the constraints are nonlinear.

In this chapter we present several production planning models, beginning with a simple single-product network model with a linear cost function and linear constraints.

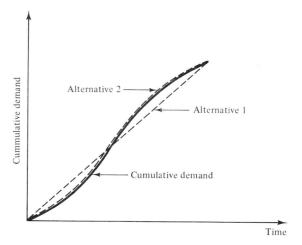

Figure 4-1. Two alternative production schedules for meeting demand.

4.2. A Simple Network Model with Linear Production and Inventory Cost

Bowman [3] developed the following production planning model, which utilizes the formulation of the transportation problem. Let

P_{ijk} = number of units produced by source i in period j to meet demand requirements in period k

C_{ijk} = marginal cost of production of a unit by source i in period j and stored to period k

B_{ij} = production capacity of source i in period j

D_k = forecasted demand requirement for the product in period k

m = number of production sources

c_R = cost per unit of regular production

c_O = cost per unit of overtime production

c_I = inventory carrying cost per unit-period

T = number of periods for the planning horizon

Z = total cost of production and inventory for all periods

The problem is then formulated as

$$\text{Minimize } Z = \sum_{i=1}^{m} \sum_{j=1}^{T} \sum_{k=1}^{T} C_{ijk} P_{ijk} \qquad (4.1)$$

subject to

$$\sum_{k=j}^{T} P_{ijk} \leq B_{ij} \qquad (i = 1, 2, \ldots, m; j = 1, 2, \ldots, T) \qquad (4.2)$$

$$\sum_{i=1}^{m} \sum_{j=1}^{k} P_{ijk} = D_k \qquad (k = 1, 2, \ldots, T) \qquad (4.3)$$

$$P_{ijk} \geq 0 \qquad (4.4)$$

The decision variable is P_{ijk}. Constraint (4.2) states that the number of units produced by source i in period j cannot exceed the capacity of the sources during that period. Constraint (4.3) states that all demand must be met on time; that is, orders cannot be backordered beyond their due dates. The cost of regular and overtime production and the inventory carrying cost comprise C_{ijk}, as will be illustrated in Example 4-1. This model attempts to balance production costs from various sources as well as inventory carrying costs such that the total cost of the production plan is minimized.

Although this is a single-product model, it should be noted that, in practice, single-product models are often used in situations where more than one product are to be produced (multiple-products). In order to formulate the multiproduct problem as a single-product model, a common denominator for the products, such as worker-hours required per unit, should be determined and treated as the unit of production. This is usually referred to as the *aggregation process*, which, after optimization, must be followed by a *disaggregation process* into the original units of production in order to achieve a master schedule. These steps are illustrated by the following example.

EXAMPLE 4-1

A chemical plant manufactures two types of products with either regular production time or through planned overtime. Products use the same equipment and are scheduled into production one at a time. Demand for the products over the next 4 months is illustrated in Table 4-1.

The initial inventory levels are 36 units of A and 220 units of B. It takes 1 plant-hour to produce a unit of product A and 0.40 plant-hour to produce a unit of B.

Table 4-1.

Month	Demand Units	
	Product A	Product B
1	100	200
2	90	190
3	110	210
4	100	200

Associated production costs are:

$$\text{cost of regular production } c_R = \$10/\text{plant-hour}$$
$$\text{cost of overtime production, } c_O = \$15/\text{plant-hour}$$
$$\text{inventory carrying cost charge, } c_I = \$4/\text{plant-hour/month}$$

Production capacities for regular time and overtime are:

$$\text{regular time} = 160 \text{ plant-hours/month}$$
$$\text{overtime} = 40 \text{ plant-hours/month}$$

Determine the production plan, in terms of plant-hours, for these products, such that the total production and inventory costs are minimized.

SOLUTION

The common denominator for these products is the plant-hour per unit. The transportation cost matrix relating sources to final demand is:

Production Period	Period of Demand 1	2	3	4	Final Inventory	Capacity (plant-hours)
Period 1						
Initial inv.	0	c_I	$2c_I$	$3c_I$	$4c_I$	124
Reg. time	c_R	$c_R + c_I$	$c_R + 2c_I$	$c_R + 3c_I$	$c_R + 4c_I$	160
Overtime	c_O	$c_O + c_I$	$c_O + 2c_I$	$c_O + 3c_I$	$c_O + 4c_I$	40
Period 2						
Reg. time	—	c_R	$c_R + c_I$	$c_R + 2c_I$	$c_R + 3c_I$	160
Overtime	—	c_O	$c_O + c_I$	$c_O + 2c_I$	$c_O + 3c_I$	40
Period 3						
Reg. time	—	—	c_R	$c_R + c_I$	$c_R + 2c_I$	160
Overtime	—	—	c_O	$c_O + c_I$	$c_O + 2c_I$	40
Period 4						
Reg. time	—	—	—	c_R	$c_R + c_I$	160
Overtime	—	—	—	c_O	$c_O + c_I$	40
Demand (plant-hours)	180	166	194	180	80	

Plant-hours of demand are computed from the demand data and plant-hour conversion factors. A planned final inventory of 80 plant-hours is considered a desirable target by management. Applying the method of linear programming yields the following solution:

Period	Demand (plant-hours)	Supply (plant-hours)
1	180	124 from initial inventory 56 from regular time of period 1
2	166	160 from regular time of period 2 6 from regular time of period 1
3	194	160 from regular time of period 3 34 from overtime of period 3
4	180	160 from regular time of period 4 20 from overtime of period 4
Final inventory	80	54 from regular time of period 1 6 from overtime of period 3 20 from overtime of period 4

The aggregate production plan, as determined above, allows the production engineer to prepare for the work-force level and overtime required during the planning horizon. For the 4-month planning period, the operating schedule is as follows:

	Planned Activity Level (plant-hours)		
Period	Regular Time	Overtime	Total Scheduled Plant-Hours
1	116	0	116
2	160	0	160
3	160	40	200
4	160	40	200

4.2.1. PRODUCTION PLAN DISAGGREGATION

We note that the production facility in Example 4-1 is a chemical plant which is shared by products A and B. Simultaneous production is not possible. Either product A or product B is produced, while retorts and mixers are thoroughly cleaned during changeover to avoid contamination between runs of different products. Therefore, one should plan the productions of each product by alternating the use of the facility between products. A disaggregation of Example 4-1 simply requires the determination of the batch sizes Q_A and Q_B that will be produced each time a changeover occurs.

As it stands, the production plan given above is a description of the production plant-hour schedule requirements, not a schedule of individual product production. In many cases this is sufficient to guide actual plant operation. For example, the demand estimates by product may represent point estimates of an uncertain process and the actual demand by product by month may be expected to differ from the forecast. In such circumstances, plant management may proceed to set hiring policies based on plant-hour production requirements and an assumption that forecast errors for individual products will be offset in the aggregate. If considerable actual variability from individual product forecasts is possible, a specific production plan by product will be less effective than a policy of dynamic, ongoing adjustment to finished-goods inventory levels. For example, one might monitor finished-goods inventory levels and produce a product when safety stock levels are reached.

In other situations, demand estimates may be very precise. Industrial

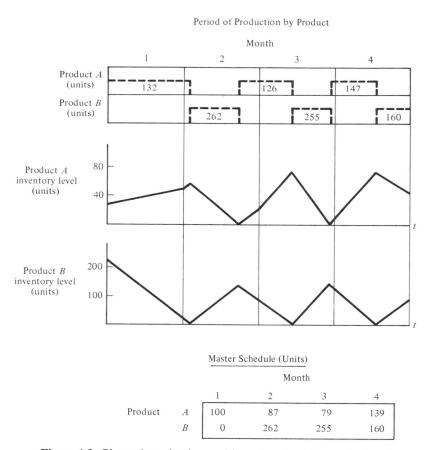

Figure 4-2. Planned production and inventory level by product and resulting master schedule.

buyers often schedule deliveries of capital goods and raw materials 6 months to a year in advance. Hence vendor companies have a relatively precise demand schedule and it is reasonable for such companies to consider a complete disaggregation of their production plan to the individual product level. When such disaggregation is done, the resulting output is called a *master schedule*, a schedule of the completion of end product production by time period.

The necessity and desirability of disaggregation is situation dependent. In Example 4-1 we will assume that the demand forecast by product is reasonably precise and we wish to disaggregate the plan. The aggregate solution has minimized two costs: the production cost and the inventory carrying cost. Given the solution, the only remaining cost to be considered is the setup cost incurred each time production is switched between products. With only two products to be considered, a simple approach to minimizing setup cost is to minimize the total number of setups scheduled over the planning horizon. One way to achieve this and thus obtain a disaggregated master schedule is to simulate production runs using the following simple decision rule: set up and produce one product until the other product's inventory runs out. At that time set up the second product and run it until the first product's inventory runs out. By continuing to alternate in this fashion, one obtains the master schedule shown in Fig. 4-2.

4.3. Linear Production Planning Models: Changing Work-Force Levels and Backlogging

In this section we extend the linear production planning model to include the costs of changing work-force levels and backlogging orders. Our objective is to determine the optimal work-force level, inventory level, and amount to be produced during any production period, such that the total cost of the production plan is minimized. These types of models can be formulated as large-scale linear programming problems. Hanssmann and Hess [6] used the linear programming modelling approach in production and employment scheduling. Models of this type are also presented in Buffa and Miller [4] and Hax [8]. We now describe a typical formulation of this variety of production planning problems. Let

D_t = forecasted demand in period t
P_t = quantity to be produced in period t (a decision variable)
C_t = unit production cost in period t (excluding labor)
I_t = on-hand inventory at the end of period t (a decision variable)
S_t = backorder quantity at the end of period t (a decision variable)

h_t = inventory carrying cost per unit held from period t to $t + 1$

W_t = backorder cost per unit short from period t to $t + 1$

L_{Rt} = regular time (worker-hours) of the work-force level in period t (a decision variable)

c_{Rt} = cost of a worker-hour of regular time during period t

L_{ot} = overtime (worker-hours) scheduled during period t (a decision variable)

c_{ot} = cost of a worker-hour of labor overtime in period t

l_t^+ = increase in work-force level in worker-hours from period $t - 1$ to t (a decision variable)

c_{lt} = cost to increase the work-force level by one worker-hour in period t

l_t^- = decrease in work-force level in worker-hours from period $t - 1$ to t

c'_{lt} = cost to decrease the work-force level by one worker-hour in period t

T = time horizon for production planning

Thus the objective function can be written to minimize the sum of labor related costs, other production costs, inventory and backlog related costs, and work-force change costs.

$$\text{Minimize } Z = \sum_{t=1}^{T} \underbrace{C_t P_t}_{\substack{\text{production} \\ \text{cost}}} + \underbrace{c_{Rt} L_{Rt} + c_{ot} L_{ot}}_{\text{labor}}$$

$$+ \underbrace{h_t I_t + W_t S_t}_{\substack{\text{inventory and} \\ \text{backorder}}} + \underbrace{c_{lt} l_t^+ + c'_{lt} l_t^-}_{\substack{\text{work-force} \\ \text{change}}} \qquad (4.5)$$

Constraints are added to ensure that the levels of variables are consistent from period to period. For example, the net inventory level at the end of any period t (NI_t) is related to the ending inventory level of the prior period $t - 1$ and the production and demand rate of the current period:

$$NI_t = NI_{t-1} + P_t - D_t \qquad (4.6)$$

$$NI_t = I_t - S_t \qquad (4.7)$$

Also, the current period's employment level is related to the prior period's employment level and the rates of increasing and decreasing the work-force level during the current period.

$$L_{Rt} = L_{R,t-1} + l_t^+ - l_t^- \qquad (4.8)$$

Overtime in any period is related to the period's scheduled production level and work-force level. Let

L_{ut} = labor undertime, that is, the worker-hours employed in excess of what is required to produce P_t

m = number of worker-hours required per unit of P_t

Then

$$L_{ot} - L_{ut} = mP_t - L_{Rt} \tag{4.9}$$

The concept of undertime may seem strange to the student of industrial engineering. *Undertime* is planned underutilization of the work force: that is, planned reductions in labor productivity. This occurs when the cost of such underutilization is less than the alternative costs of carrying additional inventory or temporarily changing the work-force level.

Finally, the nonnegativity constraint is added.

$$P_t, I_t, S_t, L_{Rt}, l_t^+, l_t^-, L_{ot}, L_{ut} \geq 0 \tag{4.10}$$

Sometimes other constraints may be desirable to add. For example, a corporate policy may put an upper limit on the maximum change in the work-force level, the scheduled overtime, and the maximum size of the backlog. Addition of such constraints to the basic model presented above is a straightforward matter.

EXAMPLE 4-2

A company produces five products in a small production facility. One product is produced at a time and machines undergo tooling changes when production is rotated among products. The forecast of monthly demand for each product is shown in Table 4-2. The initial inventory of each product and the standard worker-hours to produce a

Table 4-2. MONTHLY DEMAND OF PRODUCTS

Month	A	B	C	D	E
1	150	300	600	150	400
2	300	600	600	200	400
3	300	900	600	250	400
4	400	1500	600	400	400
5	400	600	650	350	400
6	400	600	600	350	400
7	400	100	700	350	400
8	500	0	700	500	400
9	300	0	700	550	400
10	250	0	600	475	400
11	200	0	600	450	400
12	0	0	600	400	400

Table 4-3. INITIAL INVENTORY AND STANDARD
TIMES BY PRODUCT

Product	Initial Inventory	Standard Worker-Hours per Unit
A	150	2
B	700	1
C	200	1
D	200	4
E	500	2

unit of each product are shown in Table 4-3. The following labor-related costs are estimated:

Regular time: $12 per hour

Overtime: $18 per hour

Cost of hiring: $1600 per worker

Cost of layoff: $5920 per worker

The upper limit on the work force level is 26 workers. There are currently 18 workers employed. It has been estimated that a direct labor hour of production, net of labor cost, is worth $80. The company tries to carry an average safety stock of 2 weeks of supply. The inventory carrying cost is $2.70/worker-hour per month. Assume that a worker averages 160 hours of regular time employment per month.

Develop an aggregate production plan, in terms of worker-hours, to meet the demand requirements without incurring backlogs.

SOLUTION

As stated earlier, for some production facilities worker-hours are a reasonable common denominator on which the aggregate demand could be based. The aggregate forecast of the demand in worker-hours is shown in Table 4-4.

Table 4-4. MONTHLY DEMAND (WORKER-HOURS)

Month	A	B	C	D	E	Total
1	300	300	600	600	800	2600
2	600	600	600	800	800	3400
3	600	900	600	1000	800	3900
4	800	1500	600	1600	800	5300
5	800	600	650	1400	800	4250
6	800	600	600	1400	800	4200
7	800	100	700	1400	800	3800
8	1000	0	700	2000	800	4500
9	600	0	700	2200	800	4300
10	500	0	600	1900	800	3800
11	400	0	600	1800	800	3600
12	0	0	600	1600	800	3000
Initial inventory	300	700	200	800	1000	3000

Table 4-5. AGGREGATE PRODUCTION PLAN

Month	I_t	P_t	D_t	L_{Rt}	L_{ot}	L_{ut}	I_t^+	I_t^-
1	3280	2880	2600	2880	0	0	0	0
2	3580	3700	3400	3700	0	0	820	0
3	3380	3700	3900	3700	0	0	0	0
4	2125	4045	5300	3700	345	0	0	0
5	2100	4225	4250	3700	525	0	0	0
6	1900	4000	4200	3700	300	0	0	0
7	2250	4150	3800	3700	450	0	0	0
8	2150	4400	4500	3700	700	0	0	0
9	1900	4050	4300	3700	350	0	0	0
10	1800	3700	3800	3700	0	0	0	0
11	1500	3300	3600	3700	0	400	0	0
12	1800	3300	3000	3700	0	400	0	0

Cost coefficients: Assume that the length of the period t is 1 month.

$C_t = \$80$ for each direct labor hour of production (constant for all t)
$c_{Rt} = \$12$/worker-hour
$c_{ot} = \$18$/worker-hour
$h_t = \$2.70$/worker-hour of inventory carried for 1 month
$W_t = \infty$
$c_{I_t} = 1600/160 = \$10$/worker-hour
$c'_{It} = 5920/160 = \$37$/worker-hour

In addition to Eqs. (4.5) through (4.9), L_{Rt} must be limited to 26 workers (4160 worker-hours) and I_t must be greater than the safety stock levels (2 weeks of the next period's demand).

Applying the simplex algorithm yields the aggregate production plan, shown in Table 4-5.

The purpose of this aggregate plan is to provide a broad strategy for operating the production facility based on the forecasted demand as it is known at the time of the plan. However, as the year unfolds, such plans are regularly revised on a rolling horizon. The important information to be gleaned from the plan is to recognize actions that have to be taken in the near future in order to follow the least-cost strategy. In this example, an additional five workers should be sought for hiring in the second month and overtime should be planned at the beginning of the fourth month.

4.4. Disaggregation of an Aggregate Plan

The monthly forecasted demand of the five products of Example 4-2 is given in units of products. During the aggregate planning phase, we aggregated the production in terms of worker-hours. Now we are faced with the disaggregation problem. This could have been avoided by formulating a multi-product aggregate planning model, where constraints are summed over i products and t periods. Formulations for such models are reviewed in Hax [8].

There are at least two important reasons for avoiding the multiproduct aggregate planning models in practice. The first is that integer variables must be introduced if changeover (switching) costs from one product to another are considered in the multiproduct model. Realistic size problems of the type we are considering quickly exceed the capability of a regular mixed integer programming code. Without providing information on switching, the multiproduct model does not offer any real advantages over the aggregate model.

The second important reason for avoiding the multiproduct model is that a multiproduct model solution is likely to be less accurate than that of an aggregate model. It is a common observation in practice that the errors in the forecast of demand for individual products are greater than the errors associated with the aggregation of these forecasts. The aggregation process results in offsetting the positive and negative uncorrelated individual product forecast errors. Hence the forecast of total demand (expressed in aggregate measures such as worker-hours) is likely to be much closer to the actual aggregate demand than forecasts made at the individual product level.

The disaggregation of aggregate plans into workable schedules has been studied by Hax and Meal [7], who propose a general paradigm for addressing the problem. For an example of some mathematical procedures based on the Hax and Meal approach, see Bitran and Hax [1], and Bitran, Haas, and Hax [2].

Any method for disaggregation must be highly tailored to address the specific situation of interest. In what follows, we address the problem type of Example 4-2, that is, n product disaggregation for a common production facility. This is an extension of the two-product case. However, for n products the simple runout time (time to deplete the current inventory level) rule of producing one product until it is necessary to switch over to one which is approaching its runout time does not work. We illustrate the nature of the problem in Table 4-6 by calculating the runout times for the products of Example 4-2. By following the rule of producing a product until another product runs out, it is clear that we will encounter situations in which more than one product runs out at the same time or nearly the same time (e.g., products D and E). This leads to stockouts and, in the long run, considerable additional setups.

In the section that follows we examine n-product disaggregation under

Table 4-6. INITIAL INVENTORY (WORKER-HOURS), MONTHLY DEMAND
(WORKER-HOURS/MONTH), AND RUNOUT TIMES

	Products				
	A	B	C	D	E
Initial inventory, I_0	300	700	200	800	1000
Demand in month 1	300	300	600	600	800
Demand in month 2	600	600	600	800	800
Runout time (months)	1.0	1.67	0.33	1.25	1.25

the simplified assumption of a constant production rate and constant demand rate for each product. Through the optimal disaggregation of this problem, we will illustrate some principles that can be applied in seeking relatively good solutions to the more general case of time-varying production and demand rates.

4.4.1. CONSTANT PRODUCTION AND DEMAND RATES

As in the case of two products, our objective is to minimize the number of setups per year. For ease of exposition, we consider five products whose initial inventory, demand rates, production rates, and runout times are given in Table 4-7. Inventory levels and rates are expressed in aggregate units, worker-hours for this particular case. The products have been arranged in order of increasing runout time.

We define

I_i = inventory of product i in worker-hours.
D_{it} = demand rate for product i in worker-hours per unit time in period t.
$r_i = I_i/D_{it}$ = the time at which demand exhausts the inventory of product i (runout time).
P_t = the production rate of the facility in period t.
T = the cycle time for a complete rotation of n products through the production facility.

We limit our analysis to the case where each product is produced once during a common cycle time, T. For such a case, the number of setups per year is minimized when T is maximal. We redefine the above variables to include the sequence of rotation.

$r_{[i]} = I_{[i]}/D_{[i]t}$ = the runout time of the ith product in a rotation sequence, where the bracket indicates a sequence variable.
$t_{[i]}$ = the point in time at which the production of the ith product in the rotation sequence begins.
$D_{[i]t}$ = the demand rate in period t for the ith product in the sequence.

Table 4-7. DATA FOR FIVE PRODUCTS

Product	Worker-Hours, I_0	Worker-Hours/Month D_{it}	Worker-Hours/Month P_t	Months, r_i
C	200	600	3700	0.33
A	300	600	3700	0.5
B	700	900	3700	0.78
D	800	800	3700	1.00
E	1000	800	3700	1.25

For a simple cycle, initiated at $t = 0$, T can be maximized with the following linear program:

Formulation 1

$$\text{Maximize } T \tag{4.11}$$

Subject to:

$$t_{[i]} \leq r_{[i]} \quad \forall \, [i] \tag{4.12}$$

$$(t_{[i+1]} - t_{[i]}) \frac{P_t}{D_{[i]t}} \geq T \quad [i] = [1], [2], \ldots, [n-1] \tag{4.13}$$

$$(T - t_{[n]}) \frac{P_t}{D_{[n]t}} \geq T \tag{4.14}$$

Constraint 4-12 ensures that production of product i begins prior to its runout time. Constraint 4-13 ensures that the production time allocated to product i is sufficient to cover demand over the cycle, T. Constraint 4-14 forces the cycle to begin again at T. It should be noted that this formulation assumes setup time is expended during the off shift period. It is a simple matter to include setup time in $t_{[i]}$ if that assumption does not hold.

EXAMPLE 4-3

Compute the longest simple cycle for the problem data in Table 4-7.

SOLUTION

We note that based on the runout time calculations in Table 4-7, the logical order of rotation is $C - A - B - D - E$. The linear programming problem is formulated as follows:

$$\text{Maximize } T$$

subject to

$$t_C \leq 0.33$$

$$t_A \leq 0.5$$

$$t_B \leq 0.78$$

$$t_D \leq 1.00$$

$$t_E \leq 1.25$$

$$(t_A - t_C) \frac{3700}{600} \geq T$$

$$(t_B - t_A) \frac{3700}{600} \geq T$$

$$(t_D - t_B)\frac{3700}{900} \geq T$$

$$(t_E - t_D)\frac{3700}{800} \geq T$$

$$(T - t_E)\frac{3700}{800} \geq T$$

The first constraint is unnecessary, since the production of product C will begin at $t = 0$. The last constraint indicates that the production time for the last product (product E) in the sequence, is $T - t_E$. This occurs because the second cycle must begin at T. Note also that the assumption of constant production and demand rates requires that the problem be balanced. That is,

$$\sum_i D_{it} = P_t$$

If the production rate exceeded the demand rate, inventories would grow without limit; the reverse case would result in eventual unmet demand.

The linear programming solution is

$$T^* = 1.5948$$

$$t_C = 0$$

$$t_A = 0.2586$$

$$t_B = 0.5172$$

$$t_D = 0.9052$$

$$t_E = 1.25$$

A deterministic simulation of this production cycle through the first cycle of operation is as shown in Table 4-8. Note that inventories at $t = 0$ and $t = T$ are identical; the cycle repeats every 1.5948 months. The product under production during any time period is indicated by a square around the ending inventory position of the product.

Table 4-8. DETERMINISTIC SIMULATION OF SOLUTION FOR EXAMPLE 4-3, FORMULATION 1

Time, t	Inventory Position by Product (worker-hours)					
	C	A	B	D	E	I_{total}
0	200	300	700	800	1000	3000
0.2586	1002	145	467	593	793	3000
0.5172	847	947	234	386	586	3000
0.9052	614	714	1320	76	276	3000
1.25	407	507	1010	1076	0	3000
1.5948	200	300	700	800	1000	3000

The optimal solution of Example 4-3 is determined by the runout time of product E. This is obvious since t_E has zero slack in the optimal solution, $t_E = r_E = 1.25$. For this reason, we refer to the solution of this type of problem as a *blocked maximal* solution. It is the maximal solution to the problem subject to the blocking effect of t_E.

If it were possible to shift some of the original 3000 worker-hours of inventory to product E from products that are not blocking, it would be possible to increase r_E, thus extending T. This can be accomplished by allowing some of product E to be produced prior to establishing the cycle.

We define t'_i as the time at which product i begins production prior to the beginning of the first cycle. Then the linear programming formulation that relaxes the constraint caused by product E is as follows:

Formulation 2:

$$\text{Maximize } T$$

subject to

$$t_C - t'_E \geq 0$$

$$t_C \leq 0.33$$

$$t_A \leq 0.50$$

$$t_B \leq 0.78$$

$$t_D \leq 1.00$$

$$t_E \leq 1.25 + (t_C - t'_E)\frac{3700}{800}$$

$$(t_A - t_C)\frac{3700}{600} \geq T$$

$$(t_B - t_A)\frac{3700}{600} \geq T$$

$$(t_D - t_B)\frac{3700}{900} \geq T$$

$$(t_E - t_D)\frac{3700}{800} \geq T$$

$$(t_C + T - t_E)\frac{3700}{800} \geq T$$

The first constraint ensures that product E precedes product C. The sixth constraint reflects the new runout time for product E by adding the gain in runout time from production prior to the cycle to the previous runout time of product E. The last constraint reflects the fact that the cycle begins at t_C.

Applying the method of linear programming yields the following solution:

$$T^* = 1.7158$$

$$t'_E = 0$$

$$t_C = 0.0262$$

$$t_A = 0.3044$$

$$t_B = 0.5826$$

$$t_D = 1.000$$

$$t_E = 1.3710$$

A deterministic simulation over a cycle is given in Table 4-9. The runout time for product E from the start of the cycle has been lengthened to $t_E = 1.371$. The binding constraint is now associated with the runout time of product D and product E. Note that no change occurs in the total amount of inventory that is carried (3000 worker-hours). The only change is in the distribution of that inventory among products. The process of lengthening the cycle can continue by allowing some of product D to be produced prior to establishing the cycle. If this is done, product B will then be constraining the cycle. The longest feasible cycle occurs when all products run out coincident with the start of their production. This occurs when all products are allowed to be produced prior to the first established cycle. This can be accomplished by extending the LP formulation as follows:

Formulation 3:

Table 4-9. DETERMINISTIC SIMULATION OF EXAMPLE 4-3, FORMULATION 2

	Inventory Position by Product (worker-hours)					
Time, t	*C*	*A*	*B*	*D*	*E*	I_{total}
0	200	300	700	800	1000	3000
0.0262	184	284	676	779	1077	3000
0.3044	1047	117	426	556	854	3000
0.5826	880	980	176	333	631	3000
1.0000	629	729	1345	0	297	3000
1.3710	406	506	1011	1077	0	3000
1.7420	184	284	676	779	1077	3000

Maximize T

subject to

$$t'_A \leq 0.50, \ t'_B \leq 0.78, \ t'_C \leq 0.33, \ t'_D \leq 1.0, \ t'_E \leq 1.25$$

$$t'_C - t'_A \leq 0$$

$$t'_A - t'_B \leq 0$$

$$t'_B - t'_D \leq 0$$

$$t'_D - t'_E \leq 0$$

$$t_C \leq 0.33 + \frac{3700}{600}(t'_A - t'_C)$$

$$t_A \leq 0.5 + \frac{3700}{600}(t'_B - t'_A)$$

$$t_B \leq 0.78 + \frac{3700}{900}(t'_D - t'_B)$$

$$t_D \leq 1.0 + \frac{3700}{800}(t'_E - t'_D)$$

$$t_E \leq 1.25 + \frac{3700}{800}(t'_C - t'_E)$$

$$(t_A - t_C)\frac{3700}{600} \geq T$$

$$(t_B - t_A)\frac{3700}{600} \geq T$$

$$(t_D - t_B)\frac{3700}{900} \geq T$$

$$(t_E - t_D)\frac{3700}{800} \geq T$$

$$(t_C + T - t_E)\frac{3700}{800} \geq T$$

which yields the following optimal solution:

$$T^* = 2.0404$$

$$t'_C = 0$$

$$t'_A = 0.2207$$

$$t'_B = 0.4675$$

$$t'_D = 0.8501$$

Table 4-10. DETERMINISTIC SIMULATION OF EXAMPLE 4-3, FORMULATION 3

		Inventory Position by Product (worker-hours)				
Time, t	*C*	*A*	*B*	*D*	*E*	I_{total}
0	200	300	700	800	1000	3000
0.2207	884	168	502	623	823	3000
0.4675	735	933	280	426	626	3000
0.8501	505	703	1352	120	320	3000
1.25	265	463	992	1280	0	3000
1.6911	0	199	595	927	1279	3000
2.022	1026	0	297	662	1015	3000
2.3529	827	1026	0	397	750	3000
2.8493	530	728	1389	0	353	3000
3.2904	265	464	992	1279	0	3000
3.7315	0	199	595	927	1279	3000

$$t'_E = 1.2500$$

$$t_C = 1.6911$$

$$t_A = 2.022$$

$$t_B = 2.3529$$

$$t_D = 2.8493$$

$$t_E = 3.2904$$

The deterministic simulation, shown in Table 4-10, indicates the longest feasible cycle, which begins at $t = 1.6911$ and ends at $t = 3.7315$, with each product reaching its runout point just prior to production.

Finally, it is necessary to ask whether or not it was worth adding five additional setups in order to establish the longest feasible cycle. Since the established cycle will now repeat itself continually and since we have lengthened the cycle from $T = 1.5948$ months to $T = 2.0404$ months, the average number of cycles per year has been reduced from 7.52 to 5.88. Hence the cost of the additional five setups will be returned within the first year.

4.4.2. TIME-VARYING PRODUCTION AND DEMAND RATES

Constant production and demand rates rarely exist in real situations. There may be substantial periods when both demand and production are relatively constant and for those situations the methods of Section 4.4.1 can be applied. However, more often than not, the aggregate production plan will tend to resemble the solution of Example 4-2 (i.e., time-varying production and demand). Also, in a real situation the demand forecasts are likely to be point estimates of an uncertain process. If the forecasted demand data were used to establish all the cycles over the entire time horizon, it is very unlikely that these cycles would still be appropriate after the first cycle is actually completed. By then, the difference between anticipated and actual demand would cause the planned cycles to be unworkable. For this reason, we suggest that the disaggregation of aggregate plans be treated as a dynamic process as opposed to a static one. The aggregate plan has established the plant workforce levels and production rates over time. The disaggregation problem is to establish the first production cycle and to begin to execute it. As events unfold, the cycle should be updated to reflect the actual demand and new inventory situation.

We propose a heuristic procedure whereby the blocked maximal cycle time is established for the first cycle. This is followed by production of the first product in that cycle, at which time the new inventory position and most current demand forecast is used to recompute the next blocked maximal cycle. Extension of the blocked maximal cycle can take place when the added cycle length offers considerable advantage over the additional setups required.

The following iterative procedure can be used to establish the cycle.

—*Step 1:* Choose an initial estimate of T as follows:

$$T = r_{[n]} + \frac{1}{n} r_{[n]} \tag{4.15}$$

This assumes that the actual cycle time will be somewhat more than the runout time of the last product in the sequence. Any reasonable assumption is acceptable for establishing a starting position.

—*Step 2:* Establish an initial value for the average production and demand rates as follows:

$$\bar{P}_T = \frac{1}{T} \int_0^T P_t \, dt \tag{4.16}$$

$$\bar{D}_{iT} = \frac{1}{T} \int_0^T D_{it} \, dt \tag{4.17}$$

where:

$\bar{P}_T$ = average production rate over T.

$\bar{D}_{iT}$ = average demand rate for product i over T.

—*Step 3:* Again using brackets to indicate the order of rotation, solve the linear program:

$$\text{Maximize } T \tag{4.18}$$

Subject to:

$$t_{[i]} \leq r_{[i]} \qquad \forall\, i \tag{4.19}$$

$$(t_{[i+1]} - t_{[i]}) \frac{\bar{P}_T}{\bar{D}_{[i]T}} \geq T \qquad [i] = [1], [2] \cdots [n-1] \tag{4.20}$$

Since for time-varying demand and production, $\sum_i D_{it} = P_t$ cannot be ensured, constraint (4.14) will often not be satisfied and is therefore dropped in the time-varying method.

—*Step 4:* Use the values of T and t_i from step 3 to compute a new production rate and demand rate by product. Since production and demand are time varying and since each product i must cover a cycle that begins at t_i, the computations are performed as follows:

$$\bar{P}_{[i]} = \frac{\displaystyle\int_{t_{[i]}}^{t_{[i+1]}} P_t \, dt}{t_{[i+1]} - t_{[i]}} \tag{4.21}$$

where $\bar{P}_{[i]}$ is the production rate applicable to the ith product in the sequence, which is to be produced over the interval $t_{[i+1]} - t_{[i]}$.

$$\bar{D}_{[i]} = \frac{1}{T} \int_{t_{[i]}}^{t_{[i]}+T} D_{[i]t} \, dt \tag{4.22}$$

where $\bar{D}_{[i]}$ is the demand rate applicable to the ith product in the sequence which must be satisfied over an interval $t_{[i]}$ to $t_{[i]} + T$.

—*Step 5:* Solve the linear program:

$$\text{Maximize } T \tag{4.23}$$

subject to

$$t_{[i]} \leq r_{[i]} \qquad \forall\, i \tag{4.24}$$

$$\frac{\bar{P}_{[i]}}{\bar{D}_{[i]}} (t_{[i+1]} - t_{[i]}) \geq T \qquad [i] = [1], [2], \ldots, [n-1] \tag{4.25}$$

—*Step 6:* With the new values of T and t_i, return to step 4. Continue to iterate through steps 4 and 5 until two consecutive values of T become identical.

Our experience with this algorithm indicates that convergence usually occurs quite rapidly. We illustrate with the following example.

EXAMPLE 4-4

Disaggregate the first cycle for the aggregate plan established in Example 4-2. For convenience, the initial inventory and first 3 months of production and demand are reproduced in Table 4-11.

Table 4-11. AGGREGATE PLANNING DATA (WORKER-HOURS)

| | Product Demand | | | | | Production, |
	A	B	C	D	E	P_t
Initial inventory	300	700	200	800	1000	
Month						
1	300	300	600	600	800	2880
2	600	600	600	800	800	3700
3	600	900	600	1000	800	3700
Runout time (months)	1.0	1.67	0.33	1.25	1.25	

SOLUTION

—*Step 1:* A convenient initial estimate of T is

$$T = r_{[n]} + \frac{1}{n} r_{[n]} = 1.67 + \tfrac{1}{5}(1.67) = 2$$

—*Step 2:* The estimated initial value for production and demand is

$$\bar{P}_T = \frac{1}{T} \int_0^T P_t \, dt = \frac{1}{2}\left[\int_0^1 2880 \, dt + \int_1^2 3700 \, dt \right] = 3290$$

$$\bar{D}_{iT} = \frac{1}{T} \int_0^T D_{it} \, dt$$

$$\bar{D}_A = \frac{1}{2}\left[\int_0^1 300 \, dt + \int_1^2 600 \, dt \right] = 450$$

$$\bar{D}_B = 450$$

$$\bar{D}_C = 600$$

$$\bar{D}_D = 700$$

$$\bar{D}_E = 800$$

—*Step 3:* Solve the initial LP.

Maximize T

subject to

$$t_C \leq 0.33$$

$$t_A \leq 1.0$$

$$t_D \leq 1.25$$

$$t_E \leq 1.25$$

$$t_B \leq 1.67$$

$$\frac{3290}{600}(t_A - t_C) \geq T$$

$$\frac{3290}{450}(t_D - t_A) \geq T$$

$$\frac{3290}{700}(t_E - t_D) \geq T$$

$$\frac{3290}{800}(t_B - t_E) \geq T$$

Solution:

$$T^* = 2.1538$$

$$t_C = 0$$

$$t_A = 0.3930$$

$$t_D = 0.6877$$

$$t_E = 1.1459$$

$$t_B = 1.67$$

—*Step 4:* Compute $\bar{P}_{[i]}$ and $\bar{D}_{[i]}$ using T and t_i.

$$\bar{P}_{[i]} = \frac{\displaystyle\int_{t_{[i]}}^{t_{[i+1]}} P_t \, dt}{t_{[i+1]} - t_{[i]}}$$

$$\bar{P}_C = \frac{\displaystyle\int_0^{0.3930} 2880 \, dt}{0.3930} = 2880$$

$$\bar{P}_A = \frac{\displaystyle\int_{0.3930}^{0.6877} 2880 \, dt}{0.6877 - 0.3930} = 2880$$

$$\bar{P}_D = \frac{\displaystyle\int_{0.6877}^{1} 2880 \, dt + \int_{1}^{1.1459} 3700 \, dt}{1.1459 - 0.6877} = 3141$$

$$\bar{P}_E = 3700$$

$$\bar{D}_{[i]} = \frac{\int_{t_{[i]}}^{t_{[i]} + T} D_{[i]t} \, dt}{T}$$

$$\bar{D}_C = \frac{1}{2.1538} \int_0^{2.1538} 600 \, dt = 600$$

$$\bar{D}_A = \frac{1}{2.1538} \left[\int_{0.393}^1 300 \, dt + \int_1^{2.5468} 600 \, dt \right] = 515$$

$$\bar{D}_D = 849$$

$$\bar{D}_E = 800$$

—*Step 5:* Solve the new LP.

Maximize T

subject to

$$t_C \leq 0.33$$

$$t_A \leq 1.0$$

$$t_D \leq 1.25$$

$$t_E \leq 1.25$$

$$t_B \leq 1.67$$

$$\frac{2880}{600} (t_A - t_C) \geq T$$

$$\frac{2880}{515} (t_D - t_A) \geq T$$

$$\frac{3141}{849} (t_E - t_D) \geq T$$

$$\frac{3700}{800} (t_B - t_E) \geq T$$

Solution:

$$T^* = 1.9113$$

$$t_C = 0$$

$$t_A = 0.3982$$

$$t_D = 0.7401$$

$$t_E = 1.257$$

$$t_B = 1.67$$

—*Step 6:* T^* converges in two more iterations as follows:

	Iteration Number		
	1	*2*	*3*
T^*	1.9113	1.9508	1.95
t_C	0	0	0
t_A	0.3982	0.4064	0.4054
t_D	0.7401	0.7493	0.7491
t_E	1.257	1.2482	1.2493
t_B	1.67	1.67	1.67

The third iteration yields the production time for product C that is consistent with a blocked maximal cycle time. Product C is put into production for 0.4 month, at which time the algorithm is rerun to reestablish the longest cycle. Since demand rates and production rates are time varying and actual demand is uncertain, extending the cycle using the method described for the case of constant demand and production (formations 2 and 3) is problematic. The gain in cycle length cannot be projected beyond the first cycle.

The use of the described heuristic will tend to give solutions that are consistent with the objective of minimizing setups within the constraints of the aggregate plan. During periods when the total demand rate for all products exceeds the production rate, the cycle will tend to get shorter. During periods when the production rate exceeds the total demand rate, the cycle will tend to expand.

4.5. A Quadratic Cost Model

It is argued that cost functions are usually nonlinear. A model by Holt et al. [9] begins with a quadratic cost structure obtained by approximating cost functions using both current and historical data. Through differentiation this quadratic model is converted into a system of linear equations from which one can compute optimal decision rules. These rules allow one to solve for production levels and workforce levels over the planning period. The following variables are defined:

L_t = work-force level (workers) in period t $(t = 1, 2, 3, \ldots, T)$
P_t = aggregate production rate in period t
I_t = actual aggregate net inventory at the end of period t
I_t^* = desired (ideal) aggregate net inventory at the end of period t
c_i = coefficient i, determined by curve-fitting techniques
D_t = forecasted demand for period t

Holt et al. suggest a number of typical cost functions as follows:

$$\text{regular time payroll} = c_1 L_t + c_{13} \tag{4.26}$$

where c_1 = variable cost of worker on regular time
c_{13} = fixed labor cost

$$\text{overtime cost} = c_3(P_t - c_4 L_t)^2 + c_5 P_t - c_6 L_t + c_{12} P_t L_t \tag{4.27}$$

where c_3, c_4, c_5, c_6, and c_{12} are coefficients derived from the least-squares equation:

$$\text{overtime cost} = f(L_t, P_t, L_t^2, P_t^2, L_t P_t)$$

taken from historical data. $c_3(P_t - c_4 L_t)^2$ makes the cost proportional to the workforce size. The additional terms improve the regression fit.

$$\text{inventory related cost} = c_7(I_t - I_t^*)^2$$
where
$$I_t^* = c_8 + c_9 D_t \tag{4.28}$$

It is assumed that there is an ideal end-of-period inventory level (I_t^*) which is related to aggregate demand (D_t). The coefficient is the penalty cost incurred for variations from the ideal.

$$\text{work-force level change costs} = c_2(L_t - L_{t-1} - c_{11})^2$$

Hiring and layoff costs are proportional to work-force level changes: c_2 and c_{11} are empirically derived. The complete optimization model can be formed by collecting cost terms:

$$\text{Minimize } Z = \sum_{t=1}^{T} [(c_1 - c_6)L_t + c_2(L_t - L_{t-1} - c_{11})^2$$
$$+ c_3(P_t - c_4 L_t)^2 + c_5 P_t + c_{12} P_t L_t$$
$$+ c_7(I_t - c_8 - c_9 D_t)^2 + c_{13}] \tag{4.29}$$

subject to

$$I_t = I_{t-1} + P_t - D_t \qquad t = 1, 2, \ldots, T \tag{4.30}$$

No bounds are placed on the magnitudes of P_t, L_t, and I_t; it is assumed that the quadratic costs will rule out extreme values. The first-order conditions for the objective function can be obtained by taking $(\partial Z / \partial L_r)$ $(r = 1, 2, \ldots, T - 1)$, where T is the planning horizon length.

$$\frac{\partial Z}{\partial L_r} = c_1 - c_6 + 2c_2(L_r - L_{r-1} - c_{11})$$

$$- 2c_2(L_{r+1} - L_r - c_{11}) - 2c_3 c_4(P_r - c_4 L_r) + c_{12} P_r = 0$$

which can be reduced to

$$P_r = \frac{c_{10}}{c_{14}} - c_{15} L_{r+1} + (c_{16} + 2c_{15})L_r - c_{15} L_{r-1} \qquad (4.31)$$

where

$$c_{10} = c_1 - c_6 \qquad\qquad c_{15} = \frac{2c_2}{c_{14}}$$

$$c_{14} = 2c_3 c_4 - c_{12} \qquad c_{16} = \frac{2c_3 c_4^2}{c_{14}}$$

Hence, from Eq. (4.31), the production rate (P_r) is shown to be a linear function of the work force in the current and adjacent production periods. Noting that

$$P_r = I_r - I_{r-1} + D_t$$

we can obtain the following result:

$$\frac{\partial Z}{\partial I_r} = 2c_3(P_r - c_4 L_r) - 2c_3(P_{r+1} - c_4 L_{r+1}) + c_{12} L_r$$

$$- c_{12} L_{r+1} + 2c_7(I_r - c_8 - c_9 D_r) = 0$$

where $r = 1, 2, \ldots, T - 1$

$$I_r = \frac{c_3}{c_7}(P_{r+1} - P_r) - \frac{c_{14}}{2c_7}(L_{r+1} - L_r) + c_8 + c_9 D_t \qquad (4.32)$$

Since

$$I_r = I_{r-1} + P_r - D_r$$

$$P_1 - D_1 = I_1 - I_0 = \frac{c_3}{c_7}(P_2 - P_1) - \frac{c_{14}}{2c_7}(L_2 - L_1) + c_8 + c_9 D_1 - I_0 \qquad (4.33)$$

and

$$P_r - D_r = I_r - I_{r-1} = \frac{c_3}{c_7} (P_{r+1} - 2P_r + P_{r-1})$$

$$- \frac{c_{14}}{2c_7} (L_{r+1} - 2L_r + L_{r-1}) + c_9(D_r - D_{r-1})$$

(4.34)

The production variable (P_r) can be eliminated by substituting Eq. (4.31) into Eqs. (4.33) and (4.34):

$$c_{19} L_1 - c_{20} L_2 + c_{17} L_3$$

$$= (1 - c_9)D_1 + (c_{15} + c_{17})L_0 + c_8 - \frac{c_{10}}{c_{14}} - I_0$$

(4.35)

$$- c_{21}L_1 + c_{22} L_2 - c_{21}L_3 + c_{17} L_4$$

$$= -c_9 D_1 + (1 + c_9)D_2 - c_{17} L_0 - \frac{c_{10}}{c_{14}}$$

(4.36)

$$c_{17} L_{r-2} - c_{21}L_{r-1} + c_{22} L_r - c_{21}L_{r+1} + c_{17} L_{r+2}$$

$$= -c_9 D_{r-1} + (1 + c_9)D_r - \frac{c_{10}}{c_{14}}$$

(4.37)

where

$$c_{17} = \frac{c_3 c_{15}}{c_7} \qquad c_{18} = \frac{c_3 c_{16}}{c_7} - \frac{c_{14}}{2c_7}$$

$$c_{19} = c_{16} + c_{18} + 2c_{15} + 3c_{17}$$

$$c_{20} = c_{15} + 3c_{17} + c_{18}$$

$$c_{21} = c_{15} + 4c_{17} + c_{18}$$

$$c_{22} = c_{16} + 2c_{18} + 2c_{15} + 6c_{17}$$

Equations (4.35), (4.36), and (4.37) now form a series of linear equalities of the form shown below:

$$\begin{bmatrix} c_{19} & -c_{20} & c_{17} & & & & & & \\ -c_{21} & c_{22} & -c_{21} & c_{17} & & & & & \\ c_{17} & -c_{21} & c_{22} & -c_{21} & c_{17} & & & & \\ \vdots & c_{17} & -c_{21} & c_{22} & -c_{21} & c_{17} & & & \\ \vdots & \vdots & c_{17} & -c_{21} & c_{22} & -c_{21} & c_{17} & & \\ \vdots & \vdots & \vdots & \vdots & \vdots & c_{17} & -c_{21} & c_{22} & -c_{21} & c_{17} \\ \vdots & \vdots & \vdots & \vdots & \vdots & & \cdot & \cdot & \cdot & \cdot \end{bmatrix} \begin{bmatrix} L_1 \\ L_2 \\ L_3 \\ L_4 \\ L_5 \\ \vdots \\ L_r \end{bmatrix}$$

$$
= \begin{bmatrix}
(1 + c_9)D_1 + (c_{15} + c_{17})L_0 + c_8 - (c_{10}/c_{14}) - I_0 \\
-c_9 D_1 + (1 + c_9)D_2 - c_{17}L_0 - (c_{10}/c_{14}) \\
-c_9 D_2 + (1 + c_9)D_3 - (c_{10}/c_{14}) \\
-c_9 D_3 + (1 + c_9)D_4 - (c_{10}/c_{14}) \\
-c_9 D_4 + (1 + c_9)D_5 - (c_{10}/c_{14}) \\
\vdots \quad \vdots \quad \vdots \quad \vdots \quad \vdots \quad \vdots \quad \vdots \quad \vdots \quad \vdots \quad \vdots \quad \vdots \\
-c_9 D_{r-1} + (1 + c_9)D_r - (c_{10}/c_{14}) \\
\vdots \quad \vdots \quad \vdots \quad \vdots \quad \vdots \quad \vdots \quad \vdots \quad \vdots \quad \vdots \quad \vdots \quad \vdots
\end{bmatrix}
$$

Because there is always a L_{r+1} and L_{r+2} term in each simultaneous equation, there will always be two more variables than there are equations. However, planning decisions are made and acted upon only for the first few periods. The model is run on a rolling horizon basis and revisions are made every few periods. Therefore, the equations can be solved by choosing a long planning horizon (N periods) and assuming values for L_N and L_{N-1}. Since N is large, the influence of choosing terminal conditions in the last two periods is minimized with respect to the first few periods. Having solved for L_t, we can use Eqs. (4.32), (4.33), and (4.34) to obtain values for P_t and I_t.

The approach to production planning used by Holt et al. [9] uses empirical data to estimate a quadratic cost function in L_t and P_t. This cost function is then differentiated to obtain a minimum-cost set of simultaneous equations. These equations describe the work-force level as a function of future demand and the initial conditions of the system. Solving the equations for L_t, followed by using Eqs. (4.32), (4.33), and (4.34), yields the work-force, production, and inventory levels for each period.

Although nonlinear cost functions are usually more realistic than linear cost functions, there have been few reported uses of this approach in practice. This may result from the difficulties of assessing data and fitting cost functions for the relevant cost components. Van dePanne and Bosji [16] showed that the decision rules resulting from the quadratic approach are fairly insensitive to large errors in the estimate of the cost parameters. Hence it would appear that considerable untapped advantages exist for the use of this approach in practice.

4.6. Dynamic Programming Production Planning Models

Dynamic programming has been applied to the solution of production planning problems under certain restricted assumptions. Wagner and Whitin [17] have provided such a model for the case of a fixed plus variable production cost structure, where shortages are not allowed. Zangwill [19] has extended the Wagner-Whitin algorithm to the case where backorders are allowed. We first illustrate the Wagner-Whitin algorithm, followed by Zangwill's extension to the backlog case.

4.6.1. BACKLOGGING PROHIBITED

Assume that production cost $k(P_t)$ has the following schedule:

$$k(P_t) = \begin{cases} 0, & \text{if } P_t = 0 \\ A_t + cP_t, & \text{if } P_t > 0 \end{cases}$$

where:

A_t = Fixed production cost in period t.
c = Variable unit production cost.
P_t = Production in period t.

We define the following variables:

D_t = forecasted demand in period t.
I_t = inventory at the end of period t.

Wagner and Whitin [17] have demonstrated that any optimal solution will have the properties:

$$I_{t-1}P_t = 0 \tag{4.38}$$

$$P_t = 0, D_t, D_t + D_{t+1}, D_t + D_{t+1} + D_{t+2}, \ldots, \sum_t D_t \tag{4.39}$$

Equation (4.38) states that, for any period t, either you bring inventory into the current period to service all current period demand or you bring no inventory into the period and produce the demand for the period. Equation (4.39) states that when production does take place in any period, it will be done for an entire period or combination of entire periods. The optimality of these rules has been demonstrated by Wagner and Whitin [17].

The importance of this property of an optimal solution becomes clear if one considers a simple two-period production planning problem with forecasted demand $D_1 = D_2 = 10$. If backlogging is not permitted, there exists the following 11 feasible combinations of production quantities P_t:

P_1	P_2
20	0
19	1
18	2
$\vdots$	$\vdots$
12	8
11	9
10	10

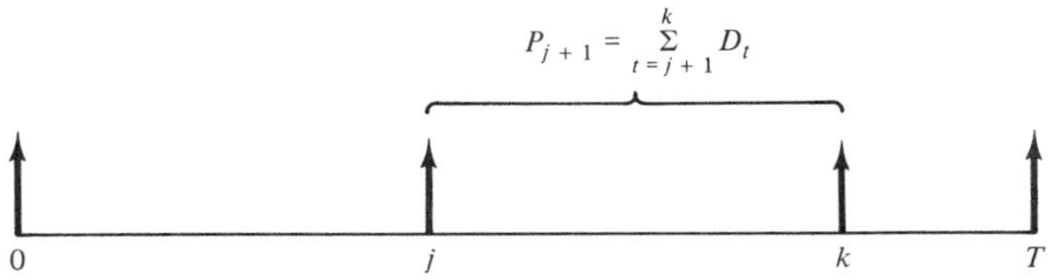

$$P_{j+1} = \sum_{t=j+1}^{k} D_t$$

Figure 4-3. Structure of multiperiod planning problem.

However,

$$I_{t-1}P_t = 0$$

implies two dominant schedules:

P_1	P_2
20	0
10	10

Hence it is only necessary to evaluate dominant schedules, greatly reducing the size of the computational effort.

The structure of the multiperiod planning situation is shown in Fig. 4-3. At the end of any period j, in which $I_j = 0$, there exists a number of possible strategies for producing to meet demand over the remainder of the planning horizon, $j + 1$, to T. Let

C_{jk} = cost of producing in period $j + 1$ to satisfy demand in $j + 1$,
$\quad j + 2, \ldots, k.$ C_{jk} includes both production cost and inventory cost.

The production cost and inventory cost over the subperiod j to k are as follows:

$$k(P_{jk}) = A_{j+1} + c(D_{j+1} + D_{j+2} + \cdots + D_k)$$

$$= A_{j+1} + cP_{j+1}$$

and

$$k(I_r) = h_r \left[P_{j+1} - \sum_{t=j+1}^{r} D_t \right]$$
$$j < r < k$$

$$k(I_{jk}) = \sum_{r=j+1}^{k-1} k(I_r)$$

where:

h_r = the holding cost for period r.
$k(P_{jk})$ = production cost for the interval j to k.
$k(I_r)$ = inventory carrying cost at the end of period r.
$k(I_{jk})$ = inventory carrying cost over the interval j to k.

Hence, the total cost of production and inventory over the period j to k can be written:

$$C_{jk} = k(P_{jk}) + k(I_{jk})$$
$$= A_{j+1} + cP_{j+1} + \sum_{t=j+1}^{k-1} h_t I_t$$

C_{jk} defines the costs associated with a subperiod j to k of a total planning horizon 0 to T. To define the dynamic programming recursion that results in a global optima, let:

$$Z_k = \min_{0 \le j \le k-1} [Z_j^* + C_{jk}] \qquad (k = 1, 2, \ldots, T)$$

At each stage of the recursion, we seek to minimize the combination of the cost of producing between two regeneration points (j and k) plus the optimal program up to j. The recursion is computed for $k = 1$ to T, where $Z_0^* = 0$

EXAMPLE 4-5

Consider a four-period problem with no initial inventory and the following data:

Period, t	Demand Forecast, D_t	Setup Cost, A_t	Unit Variable Cost, c_t	Unit Holding Cost per Period, h_t
1	10	$20	$3	$2
2	30	40	3	2
3	30	30	3	1
4	20	50	3	1

SOLUTION

$$Z_1 = C_{01} = A_1 + c_1(D_1) = 20 + (3)(10) = 50^*$$

$$Z_2 = \begin{cases} Z_0^* + C_{02} = A_1 + c_1(D_1 + D_2) + h_1(D_2) \\ \qquad = 20 + 3(40) + 2(30) = 200 \\ Z_1^* + C_{12} = Z_1^* + A_2 + c_2(D_2) \\ \qquad = 50 + 40 + 3(30) = 180^* \end{cases}$$

$$Z_3 = \begin{cases} Z_0^* + C_{03} = 20 + 3(70) + 2(60) + 2(30) = 410 \\ Z_1^* + C_{13} = 50 + 40 + 3(60) + 2(30) = 330 \\ Z_2^* + C_{23} = 180 + 30 + 3(30) = 300* \end{cases}$$

$$Z_4 = \begin{cases} Z_0^* + C_{04} = 20 + 3(90) + 2(80) + 2(50) + 1(20) = 570 \\ Z_1^* + C_{14} = 50 + 40 + 3(80) + 2(50) + 1(20) = 450 \\ Z_2^* + C_{24} = 180 + 30 + 3(50) + 1(20) = 380* \\ Z_3^* + C_{34} = 300 + 50 + 3(20) = 410 \end{cases}$$

The production schedule is found by tracing the solution backward:

$$Z_4^* = 380 \Rightarrow \text{produce 50 units in period 3 for periods 3 and 4}$$
$$Z_2^* = 180 \Rightarrow \text{produce 30 units in period 2 for period 2}$$
$$Z_1^* = \ \ 50 \Rightarrow \text{produce 10 units in period 1 for period 1.}$$

The optimal solution is as follows:

Period	Production (units)
1	10
2	30
3	50
4	0

4.6.2 BACKLOGGING ALLOWED

In a postscript to the methods of section 4.6.1, Wagner [18] demonstrated that the Wagner-Whitin algorithm is also applicable to situations in which the cost curves differ from period to period but have non-increasing marginal cost as a function of output. This applies to the case where costs are concave and includes the special case where:

$$k(P_t) = \begin{cases} 0, & \text{if } P_t = 0 \\ A_t + c_t P_t, & \text{if } P_t > 0 \end{cases}$$

In other words, c_t does not have to be constant for all t.

Building on this, Zangwill [19] extended the Wagner-Whitin algorithm to the case where backlogging is permitted. The production decision situation is shown in Fig. 4-4. C_{jk} is the minimum cost of producing in exactly one of the periods $j + 1, j + 2, \ldots, k$ to satisfy the demand over the periods $j + 1$ to k. If production occurs during period l, backorders are accumulated from $j + 1$ to l.

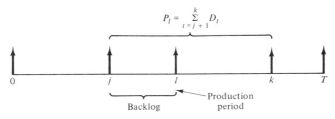

Figure 4-4. Structure of multiperiod planning with back-logging.

Production in l immediately satisfies those backorders and also provides the inventory to satisfy demand from $l + 1$ to k. More precisely since

$$k(P_{jk}) = A_\ell + c_\ell(D_{j+1} + D_{j+2} + \cdots D_k)$$
$$= A_\ell + c_\ell P_\ell$$

where $P_\ell = \sum_{t=j+1}^{k} D_t$. Inventory and backlogging costs are

$$k(I_{jk}) = \sum_{r=\ell}^{k-1} k(I_r) = \sum_{t=\ell}^{k-1} h_t I_t$$
$$k(S_{jk}) = \sum_{r=j+1}^{\ell-1} k(S_r) = \sum_{t=j+1}^{\ell-1} W_t S_t$$

where $S_t =$ backlog at the end of period t
$W_t =$ backlog penalty cost per unit backlogged

Hence,

$$C_{jk} = \min_{j+1 \leqslant \ell \leqslant k} \left[A_\ell + c_\ell P_\ell + \sum_{t=j+1}^{\ell-1} W_t S_t + \sum_{t=\ell}^{k-1} h_t I_t \right]$$

and the dynamic programming recursion is again:

$$Z_k = \min_{0 \leqslant j \leqslant k-1} [Z_j + C_{jk}]$$

EXAMPLE 4-6

Consider the problem of Example 4-5 with the following backlog costs:

Period	W_t
1	$1
2	1
3	2
4	2

SOLUTION

$$C_{01} = \quad A_1 + c_1(D_1) = 20 + 3(10) = 50$$

$$C_{02} = \begin{cases} A_1 + c_1(D_1 + D_2) + h_1(D_2) = 20 + 3(40) + 2(30) = 200 \\ A_2 + c_2(D_1 + D_2) + W_1(D_1) = 40 + 4(40) + 1(10) = 210 \end{cases}$$

$$C_{03} = \begin{cases} A_1 + c_1(D_1 + D_2 + D_3) + h_1(D_2 + D_3) + h_2(D_3) \\ \quad = 20 + 3(70) + 2(60) + 2(30) = 470 \\ A_2 + c_2(D_1 + D_2 + D_3) + h_2(D_3) + W_1(D_1) \\ \quad = 40 + 4(70) + 2(30) + 1(10) = 390 \\ A_3 + c_3(D_1 + D_2 + D_3) + W_1(D_1) + W_2(D_1 + D_2) \\ \quad = 30 + 4(70) + 1(10) + 1(40) = 360 \end{cases}$$

$$C_{04} = \begin{cases} A_1 + c_1(D_1 + D_2 + D_3 + D_4) + h_1(D_2 + D_3 + D_4) + h_2(D_3 + D_4) + h_3(D_4) \\ \quad = 20 + 3(90) + 2(80) + 2(50) + 1(20) = 570 \\ A_2 + c_2(D_1 + D_2 + D_3 + D_4) + h_2(D_3 + D_4) + h_3(D_4) + W_1(D_1) \\ \quad = 40 + 4(90) + 2(50) + 1(20) + 1(10) = 530 \\ A_3 + c_3(D_1 + D_2 + D_3 + D_4) + h_3(D_4) + W_1(D_1) + W_2(D_1 + D_2) \\ \quad = 30 + 4(90) + 1(20) + 1(10) + 1(40) = 460 \\ A_4 + c_4(D_1 + D_2 + D_3 + D_4) + W_1(D_1) + W_2(D_1 + D_2) + W_3(D_1 + D_2 + D_3) \\ \quad = 50 + 5(90) + 1(10) + 1(40) + 2(70) = 690 \end{cases}$$

The subsets C_{12}, C_{13}, and C_{14} are

$$C_{12} = \quad A_2 + c_2(D_2) = 40 + 4(30) = 160$$

$$C_{13} = \begin{cases} A_2 + c_2(D_2 + D_3) + h_2 D_3 \\ \quad = 40 + 4(60) + 2(30) = 340 \\ A_3 + c_3(D_2 + D_3) + W_2 D_2 \\ \quad = 30 + 4(60) + 1(30) = 300 \end{cases}$$

$$C_{14} = \begin{cases} A_2 + c_2(D_2 + D_3 + D_4) + h_2(D_3 + D_4) + h_3 D_4 \\ \quad = 40 + 4(80) + 2(50) + 4(20) = 540 \\ A_3 + c_3(D_2 + D_3 + D_4) + h_3 D_4 + W_2 D_2 \\ \quad = 30 + 4(80) + 1(20) + 1(30) = 400 \\ A_4 + c_4(D_2 + D_3 + D_4) + W_2 D_2 + W_3(D_2 + D_3) \\ \quad = 50 + 5(80) + 1(30) + 2(60) = 600 \end{cases}$$

The subsets C_{23} and C_{24} are

$$C_{23} = A_3 + c_3 D_3 = 30 + 4(30) = 150$$

$$C_{24} = \begin{cases} A_3 + c_3(D_3 + D_4) + h_3 D_4 \\ \quad = 30 + 4(50) + 1(20) = 250 \\ A_4 + c_4(D_3 + D_4) + W_3(D_3) \\ \quad = 50 + 5(50) + 1(30) = 330 \end{cases}$$

The subset C_{34} is

$$C_{34} = A_4 + c_4 D_4 = 50 + 5(20) = 150$$

SUMMARY OF RESULTS

Last Regeneration Point, j	Next Regeneration Point, k	Period of Production, t				$t^*(j, k)$	C_{jk}
		1	2	3	4		
0	1	50				1	50
	2	200	210			1	200
	3	470	390	360		3	360
	4	570	530	460	690	3	460
1	2		160			2	160
	3		340	300		3	300
	4		540	400	600	3	400
2	3			150		3	150
	4			250	330	3	250
3	4				150	4	150

The summary of results indicates, for each possible regeneration point combination (i and j), the optimal period of production. For example, if C_{03} were an optimal subperiod, then production in period 3 would be optimal for that subperiod. However, the summary does not indicate which of the subperiods are optimal regeneration subperiods. To evaluate this, one applies the recursion

$$Z_k = \min_{0 \leqslant j \leqslant k-1} [Z_j + C_{jk}]$$

$$Z_1 = Z_0 + C_{01}^* = 0 + 50 = 50$$

$$Z_2 = \begin{cases} Z_0 + C_{02}^* = 0 + 200 = 200 \\ Z_1 + C_{12}^* = 50 + 160 = 210 \end{cases}$$

$$Z_3 = \begin{cases} Z_0 + C_{03}^* = 0 + 360 = 360 \\ Z_1 + C_{13}^* = 50 + 300 = 350 \\ Z_2 + C_{23}^* = 200 + 150 = 350 \end{cases}$$

$$Z_4 = \begin{cases} Z_0 + C^*_{04} = 0 + 460 = 460 \\ Z_1 + C^*_{14} = 50 + 400 = 450 \\ Z_2 + C^*_{24} = 200 + 250 = 450 \\ Z_3 + C^*_{34} = 350 + 150 = 500 \end{cases}$$

Summary Z_k

		1	2	3	4
	0	50*	200*	360	460
	1		210	350	450
j	2			350	450*
	3				500
	Z_k	50	200	350	450

The optimal regeneration points and production program is as follows:

Period	Production (units)
1	40
2	0
3	50
4	0

4.7. Summary

The purpose of production planning is to develop an overall strategy for production over the planning horizon. Production plans are usually aggregate in nature and producers initiate actions based only on the first few periods of the plan. Plans are updated regularly as the demand and inventory situation changes.

A few approaches to developing an aggregate plan have been illustrated. Linear cost modeling is by far the most popular due to the availability of good linear programming codes to handle large problems. Methods have been developed for quadratic and other nonlinear cost functions, but their reported use in practice is minor.

A difficult problem in aggregate planning is that of disaggregating the optimal plan. Disaggregation has been accomplished when a production sequence and lot sizes have been established for each of the products in the plan. Examples of the disaggregation process in product layouts which produce multiple products were illustrated.

The more difficult problem of plan disaggregation in job shops and other discrete parts production facilities has not been specifically addressed through an example. In practice, the planning function for this kind of production environment is carried out through the method of *materials requirements planning*. This is the subject of the next chapter.

PROBLEMS

4-1. A producer is considering either producing or procuring a certain part. The relevant data are as follows:

	Period			
	1	*2*	*3*	*4*
Production capacity				
In-house	30	40	60	60
Outside vendor	60	60	60	60
Unit cost				
In-house	20	20	21	22
Outside vendor	26	26	26	26
Demand	60	80	100	120

Initial inventory is 20 units and the holding cost is $5 per unit per period. What is the optimal aggregate plan?

4.2. In Prob. 4-1 the outside vendor is willing to give discounts based on the size of the order placed. Furthermore, the order will be delivered in parts over time based upon predetermined delivery schedule. Payment for the quantities delivered is made only at time of delivery. The discount schedule is an *all-units* discount, as follows:

Order Size (units)	Price per Unit
0–60	$26
61–90	24
91–120	22
121–180	20
> 180	20

What is the optimal production plan?

4-3. A jelly manufacturer bottles three flavors of jelly on the same automatic filling machinery. Demand for each brand over the next six weeks is as follows:

	Units Demanded ($\times 10^3$)		
Week	*Grape*	*Strawberry*	*Apricot*
1	100	50	100
2	100	80	80
3	200	100	150
4	150	120	80
5	200	50	80
6	200	100	100
Initial inventory	200	100	150

The filling machinery is operated for 40 regular time hours per week maximum and overtime is limited to 20% of the scheduled regular time hours.

Due to the difference in viscosity among the three jellies, the filling rates are different. The standard number of jars filled per hour is as follows:

Jelly	Jars Filled per Hour at Standard
Grape	9800
Strawberry	7200
Apricot	6900

Cost per regular time and overtime operating hours are as follows:

Regular time:	$200
Overtime:	400

The carrying cost rate the firm applies to a dollar of inventory is 40% per annum. What is the optimal aggregate production plan?

4-4. Develop a master production schedule from the answer to Prob. 4-3. Assume that all changeovers are scheduled as overtime on the off-shift.

4-5. A manufacturer has three manufacturing cells, each charged with the production of five different products. Demand, initial inventory, and standard hours per unit for each cell and for each product within the cell are given in the tables below. The following cost data apply:

Cost of regular time/worker-hour:	$ 12/hour
Cost of overtime/worker-hour:	18/hour
Cost of hiring:	1600/worker
Cost of layoff:	5920/worker
Inventory carrying charge:	2.70/worker-hour of production carried per month

Other applicable conditions are as follows:

Cell	Current Employment	Maximum Employment	Maximum Overtime
I	18	26	40% of regular time
II	32	50	40% of regular time
III	20	25	40% of regular time

Safety stock levels, or inventory, will not be lower than 2 weeks of demand in total.

(a) Develop an aggregate plan for cell I only.

(b) Develop an aggregate plan for the entire plant. Assume that workers are allowed to be transferred between cells from period to period. Due to learning during relocation, there is a cost to transfer of $240 per worker.

Monthly demand by product is given in Table 4-11.

Table 4-11. UNITS DEMANDED

	Cell I					Cell II						Cell III			
Month	A	B	C	D	E	F	G	H	I	J	K	L	M	N	O
1	150	300	600	150	400	1000	300	300	150	1000	1000	100	300	150	600
2	300	600	600	200	400	800	300	400	300	800	800	100	600	300	600
3	300	900	600	250	400	600	300	500	450	600	600	100	900	300	600
4	400	1500	600	400	400	600	300	800	750	600	600	100	1500	400	600
5	400	600	650	350	400	600	300	700	300	600	600	100	600	400	650
6	400	600	600	350	400	600	300	700	300	600	600	100	600	400	600
7	400	100	700	350	400	700	300	700	50	700	700	100	100	400	700
8	500	0	700	500	400	900	300	1000	0	900	900	100	0	500	700
9	300	0	700	550	400	1500	300	1100	0	1500	1500	100	0	300	700
10	250	0	600	475	400	1800	300	950	0	1800	1800	100	0	250	600
11	200	0	600	450	400	1500	300	900	0	1500	1500	100	0	200	600
12	0	0	600	400	400	1000	300	800	0	1000	1000	100	0	0	600
Initial inventory	150	700	200	200	500	500	400	300	100	700	400	50	200	300	600
Standard hours per unit	2	1	1	4	2	2	1	3	1	2	2	3	1	1	1

4-6. Disaggregate the aggregate plan of Prob. 4-5(a). Assume that the cell is tooled to work on only one product at a time.

4-7. Solve the following Wagner–Whitin problem.

	Period		
	1	*2*	*3*
Demand (units)	100	160	150
Fixed setup cost	$60	$70	$50
Variable unit production cost	$3.25	$3.50	$4.00
End of period unit inventory holding cost	$0.80	$0.80	$1.00
Initial inventory = 40 units			

4-8. What is the optimal production plan for the following problem?

COST PER UNIT

	Period			
Range	*1*	*2*	*3*	*4*
$0 < X \leq 10$	7	5	7	8
$10 < X \leq 20$	8	7	7	9
$20 < X \leq 30$	9	9	8	10
$30 < X \leq 40$	10	10	12	12

Demand and inventory holding costs are as follows:

Period	*Demand*	*Inventory Holding Cost*
1	26	1
2	32	2
3	40	2
4	32	3

4-9. Solve the following five-period production planning problem with backlogging.

Period	D_t	A_t	c_t	h_t	W_t
1	20	80	18	1	4
2	30	120	20	1	4
3	25	90	18	1	4
4	40	100	16	2	4
5	10	100	16	2	4

REFERENCES

[1] BITRAN, GABRIEL, R., AND ARNOLD C. HAX, "On the Design of Hierarchical Production Planning Systems," *Decision Sciences*, 8, 1977.

[2] BITRAN, GABRIEL R., ELIZABETH A. HAAS, AND ARNOLD C. HAX, "Hierarchical Production Planning: A Single Stage System," *Operations Research*, Vol. 29, No. 4, July/Aug., 1981.

[3] BOWMAN, EDWARD, H., "Production Scheduling by the Transportation Method of Linear Programming," *Operations Research*, 3, no. 1, 1956.

[4] BUFFA, ELWOOD AND J. G. MILLER, *Production Inventory Systems: Planning and Control,* Richard D. Irwin, 1979.

[5] HANSSMAN, FRED, *Operations Research in Production and Inventory Control* New York: John Wiley and Sons, 1962.

[6] HANSSMAN, FRED, AND S. W. HESS, "A Linear Programming Approach to Production and Employment Scheduling," *Management Technology*, January, 1960.

[7] HAX, ARNOLD C., AND HARLAN C. MEAL, "Hierarchical Integration of Production Planning and Scheduling," *TIMS Studies in Management Science*, 1, Logistics, Murray Geisler, ed., January 1975.

[8] HAX, ARNOLD C., "Aggregate Production Planning," *Handbook of Operations Research: Models and Applications*, Salah E. Elmaghraby, and Joseph J. Moder, eds. New York, N.Y.: Van Nostrand Reinhold Co., 1978.

[9] HOLT, CHARLES C., J. FRANCO MODIGLIANI, JOHN F. MUTH, AND HERBERT A. SIMON, *Planning Production, Inventories, and Workforce*. Englewood Cliffs, N.J.: Prentice-Hall, Inc., 1960.

[10] JOHNSON, LYNWOOD A. AND DOUGLAS C. MONTGOMERY, Operations Research in Production Planning, Scheduling and Inventory Control. New York, N.Y., John Wiley & Sons, Inc., 1974.

[11] KHOSHNEVIS, BEHROKH AND PHILIP M. WOLFE, "An Aggregate Production Planning Model Incorporating Dynamic Productivity," *IIE Transactions*, Vol. 15, No. 2, June 1983.

[12] KHOSHNEVIS, BEHROKH, PHILIP M. WOLFE, AND M. PALMER TERRELL, "Aggregate Planning Models Incorporating Productivity—An Overview," *International Journal of Production Research*, Vol. 20, No. 5, Sept./Oct., 1982.

[13] KLEIN, MORTON, "A Transportation Model for Production Planning with Convex Costs," *IIE Transactions*, Vol. 15, No. 3, Sept. 1983.

[14] RAPP, BIRGES AND ANDERS THORSTENSON, "Product Variants and Turnover Rate," *International Journal of Production Research*, Vol. 21, No. 4, July/Aug. 1983.

[15] SAAD, GERMAINE H., "An Overview of Production Planning Models: Structural Classification and Empirical Assessment," *International Journal of Production Research*, Vol. 20 No. 1, Jan./Feb. 1982.

[16] VAN DEPANNE, C., AND P. BOSJI, "Sensitivity Analysis of Lost Coefficient Estimates: The Case of Linear Decision Rules for Employment and Production," *Management Science*, 9, 1962.

[17] WAGNER, HARVEY M., AND THOMAS M. WHITIN, "A Dynamic Version of the Economic Lot Size Model", *Management Science,* 5, 1958.

[18] WAGNER, HARVEY M., "A Postscript to 'Dynamic Problems of the Firm'," Naval Research Logistics Quarterly, Vol. 7, March, 1960.

[19] ZANGWILL, WILLARD I., "A Deterministic Multiperiod Production Scheduling Model with Backlogging," *Management Science*, 13, 1, 1966.

[20] ZANGWILL, WILLARD I., "Production Smoothing of Economic Lot Sizes with Non-Decreasing Requirements," *Management Sciences*, 13, 3, 1966.

MATERIAL REQUIREMENTS PLANNING

5.1. Introduction

The methods of aggregate planning as described in Chapter 4 do not account for the detailed timing of flows within the production system. In production systems designed as product layouts, or flow shops, the disaggregation procedures that were described in Chapter 4 will usually suffice to give workable production plans by product. This is true because production lead times are usually short in such systems and there is a simple relationship between the technological ordering of production requirements for manufacturing the product and the physical layout of the production system. However, in production systems designed as process layouts, or job shops, the flow of the product is more complex. Component workpieces move independently between separate departments and share scarce common resources. Since workpieces must be transported between departments and often are queued before being scheduled on a machine within a department, the production lead times are very long, often in the order of months. Consequently, the control of product flow and the utilization of resources in the process layout system is a more difficult and challenging planning problem.

In practice, ensuring that production resources have relatively high utilization is often accomplished by building buffer stocks of work-in-process inventory between departments and between operations. It is a common problem in discrete parts manufacturing using a process layout that the investment in work-in-process inventory is very high.

5

The method of *material requirements planning* (MRP) was developed specifically for the purpose of dealing with the complexity of these timing and inventory relationships in the discrete parts manufacturing environment. Fundamental to the MRP approach is the distinction that is made between *independent* and *dependent* demand as it affects inventory and production decisions. In Chapter 3 we examined several inventory models based on assumptions of either constant demand or statistical distributions of demand. In practice, it is rare to find situations in which demand is constant, although this may be a good approximation where the variance to mean ratio of demand is quite small. More often than not, demand for an item comes in lumps; that is, some units may be demanded during one period of time and then there will be no demand for awhile. This is especially true within a discrete parts production facility. Demand for a particular blank of raw material stock from which a part is to be machined occurs only when that part goes into production. In a discrete parts production facility, in which there are typically 20,000 to 100,000 component parts, the demand for that particular blank, will be infrequent. Hence inventorying that blank based on a continuous demand model is inappropriate.

A statistical inventory model may be appropriate to describe final demand for the end products of the firm. End product demand is said to be *independent* demand, since it originates from independent sources outside the production system. However, the demand for subassemblies, component parts, and raw material stock is derived from the planned production levels of the end products. Once production planning determines the weekly master production schedule for end products, the requirements for subassemblies, components, and raw stock items related to those end products can be simply computed. For this reason, it is said that demand for these items are *dependent* on the planned production of final products.

The manufacturing routing sheets and product bills of materials describe the departmental routings and production times to manufacture the subassemblies and components. Using these data bases in conjunction with a schedule of end product requirements, it is possible to compute the timing of production for each component to meet the given end product schedule. This, in effect, is what an MRP system does. In other words, given a master schedule of end (or final) product, MRP computes the timing of all the subassembly, component, and raw material production and purchasing activities required over the specified production horizon to meet the master schedule of the end product. Moreover, it does so in such a way as to attempt to minimize work-in-process inventory.

The methods of MRP are not new. However, until recently, it has been economically impractical to employ these methods in any nontrivial production situation. The dramatic reduction in computer cost over the last 20 years has changed that circumstance. As we shall see, operating an MRP system requires massive data storage, retrieval and computational capabilities which can only be accommodated by computers. This will be discussed later in this chapter when the software structure of an MRP system is described. First, it is necessary to investigate the product structure relationships that MRP takes advantage of in solving the production planning problem.

5.2. Parts Explosion Requirements

Once a master schedule of end product production is determined, production planning must address the problem of producing subassemblies and components. Components are produced or purchased from which subassemblies and final assemblies are eventually made. Therefore, when a master schedule is proposed, the first problem is to convert it efficiently into the requirements for purchased and manufactured parts. This is referred to as performing a *parts explosion*.

To present a formal description of the problem, we classify items to be produced into three categories: *end items*, *subassemblies*, and *parts*. They are referred to by the notation e, s, and p, respectively. Let the bill of materials for an end product, i, be defined by the row vector

$$B^i = (b_{i1}, b_{i2}, \ldots, b_{ij})$$

where b_{ij} is the number of units of item j required to make directly 1 unit of

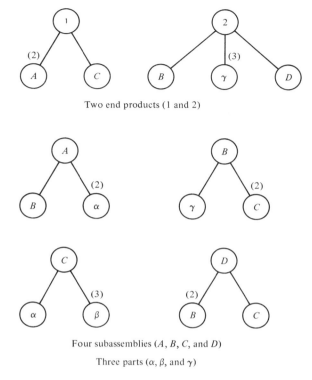

Two end products (1 and 2)

Four subassemblies (*A*, *B*, *C*, and *D*)

Three parts (α, β, and γ)

Figure 5-1. Product structure for two hypothetical end products.

item *i*. Then

$$
B = \begin{pmatrix} B^1 \\ B^2 \\ \vdots \\ B^n \end{pmatrix} = \text{bill of materials (BOM) matrix}
$$

For example, consider the bill of materials structure for end products 1 and 2 shown in Fig. 5-1. The numbers in parentheses indicate the quantity of an item required; otherwise, the quantity is 1.

The bill of materials matrix structure that captures the information in Fig. 5-1 is shown in Table 5-1. The rows of Table 5-1 are the *how constructed files* of the BOM; reading across, one can determine exactly what is required to go directly into an end product or subassembly. For example, product 1 requires 2 units of *A* and 1 unit of *C*. This corresponds to the organization in Fig. 5-1. The columns of Table 5-1 are the *how used files*. For example, *C* is used directly in the production of product 1 and subassemblies *D* and *B*.

Table 5-1. BILL OF MATERIALS MATRIX

	End Product		Subassemblies				Parts (Components)		
	1	*2*	*A*	*D*	*B*	*C*	*γ*	*α*	*β*
1			2			1			
2				1	1		3		
A					1			2	
D					2	1			
B						2	1		
C								1	3
γ									
α									
β									

An important characteristic of the organization of data in an MRP system is indicated by the upper triangularity of the B matrix in Table 5-1. In an MRP system, it is common for subassemblies and components to be numbered by levels. These level numbers describe the relative position of assemblies and parts to one another and to end products. Level numbers are assigned on the basis of the maximum number of stages of assembly required to get the subassembly or the part into an end product. Later, when subassembly and component production is planned, the computation of requirements proceeds sequentially from the highest level to the lowest level.

As an example of the level numbering system, consider Fig. 5-1 in relation to Table 5-2. From Fig. 5-1 it can be seen that subassembly A goes directly into product 1 and it is not used anywhere else. Hence, subassembly A is one level removed from an end product and is assigned to level 1, as shown in Table 5-2. By similar reasoning, it can be shown that subassembly D is a level 1 subassembly.

Subassembly B goes directly into end product 2. However, subassembly B is also used to produce level 1 subassemblies A and D. Hence the maximum number of steps of assembly required to reach the end product is 2: first, B is assembled into A, then A is assembled into 1. Hence B is a level 2 item.

By continuing the same reasoning process, level assignments are made as shown in Table 5-2. These level assignments correspond to the ordering of items within categories in matrix B given in Table 5-1.

The Bill of Material Matrix has been used as the basis for developing methods of computing total production requirements generated by the

Table 5-2. MRP LEVEL ASSIGNMENTS

Level				
0	*1*	*2*	*3*	*4*
1	A	B	C	α
2	D		γ	β

demand for end products. The methods, explained in the following sections, are based largely on the work contained in References [6, 7, 8, 14, 15].

5.3. Computing Direct Dependent Demand

Using the structure of the B matrix and the level assignments of Table 5-2, one can compute direct dependent demand at any level. Let

d_n = vector of demand at level n
$dd(n)$ = vector of dependent demand directly resulting from demand at level n

Then

$$dd(n) = d_n \times B \qquad (5.1)$$

EXAMPLE 5-1

Assuming the product structure of Fig. 5-1, compute the direct dependent demand which results from the following level 0 demand:

Item 1: 100 units

Item 2: 200 units

SOLUTION

$dd(0) = d_0 \times B$

$$dd(0) = (100 \quad 200 \quad 0 \quad 0 \quad 0 \quad 0 \quad 0 \quad 0 \quad 0) \begin{bmatrix} 0 & 0 & 2 & 0 & 0 & 1 & 0 & 0 & 0 \\ 0 & 0 & 0 & 1 & 1 & 0 & 3 & 0 & 0 \\ 0 & 0 & 0 & 0 & 1 & 0 & 0 & 2 & 0 \\ 0 & 0 & 0 & 0 & 2 & 1 & 0 & 0 & 0 \\ 0 & 0 & 0 & 0 & 0 & 2 & 1 & 0 & 0 \\ 0 & 0 & 0 & 0 & 0 & 0 & 0 & 1 & 3 \\ 0 & 0 & 0 & 0 & 0 & 0 & 0 & 0 & 0 \\ 0 & 0 & 0 & 0 & 0 & 0 & 0 & 0 & 0 \\ 0 & 0 & 0 & 0 & 0 & 0 & 0 & 0 & 0 \end{bmatrix}$$

$dd(0) = (0 \quad 0 \quad 200 \quad 200 \quad 200 \quad 100 \quad 600 \quad 0 \quad 0)$

The number of subassemblies and components required for the production of 100 units of end product 1 and 200 units of end product 2 are:

Item	Demand
A	200
D	200
B	200
C	100
γ	600

In Example 5-1, we computed the first order requirements for assembling the demanded end products. Thus, 600 units of component γ are required because we wish to assemble 200 units of end product 2 and each unit requires 3 units of γ. However, there are second order requirements for γ that are generated by this demand. For example, end product 2 requires one unit of B and each B requires one unit of γ. Hence, to completely specify dependent demand for γ, it is necessary to compute total requirements.

5.4. Computing Total Requirements

Using the bill of materials matrix B and the planned production of end products, it is of interest to compute the total requirements of all subassembly and component production required to meet end product demand. With regard to this, it is important to note that component parts and subassemblies enter the final product both directly and indirectly. In order to account for the complete requirements of an nth-level component, one must sum its relationship to subassemblies and final products. In this regard, we can attach a special interpretation to the structure of the B matrix.

If B is an $n \times n$ triangular matrix, then $B^k = 0$ for any $k \geq n$. The proof of this is intuitively obvious. By definition, the diagonal of B is all zeros. By matrix multiplication, B^2 must have a zero diagonal above the main diagonal, B^3 will have two such zero diagonals above the main diagonal, and so on. Therefore, B^n must be a zero matrix. In order to interpret these powers of B, we define

$$b_{ij}^2 = i, j\text{th entry of } B^2$$

By matrix multiplication,

$$b_{ij}^2 = \sum_{k=1}^{i-1} b_{ik} b_{kj} = b_{i1}b_{1j} + b_{i2}b_{2j} + \cdots + b_{i,\,i-1}b_{i-1,\,j}$$

where the sum is terminated at $i - 1$ because b_{ij} is on or below the diagonal when $k \geq i$. We note that for b_{ij}^2 to be nonzero, one or more of the products $b_{ik}b_{kj}$ must be nonzero. These products are the two-stage requirements of i for j through k. We conclude that i has a two-stage requirement for j and that B^2 gives the number of two-stage requirements for items.

As an example, consider the $(1, 8)$ entry of the B^2 matrix, that is, the two-stage requirement for component α in product 1. Then

$$b_{1,\,8}^2 = \sum_k b_{1k} b_{k8} = 0 + 0 + 2(2) + 0 + 0 + 1(1) + 0 + 0 = 5$$

This result occurs because end product 1 requires two A and one C. Since each

Table 5-3. B^2 MATRIX

	1	2	A	D	B	C	γ	α	β
1					2			5	3
2					2	3	1		
A					2	1			
D					4	2		1	3
B								2	6
C									
γ									
α									
β									

A requires two α and each C requires one α, the two-stage requirement for α is 5. The complete two-stage requirement matrix for B is given in Table 5-3.

The matrix of total requirements can be thought of as the sum of the n-stage requirements matrices. It can be computed directly as follows: Let

R = total requirements matrix

$$R = \begin{bmatrix} R^1 \\ R^2 \\ R^3 \\ \vdots \\ R^n \end{bmatrix} \text{ where } R^i \text{ is a row vector of total requirements for item } i$$

$$R^i = (r_{i1}, r_{i2}, \ldots, r_{ij})$$

where r_{ij} = total number of units of item j required to produce one unit of item i; this includes those units of j entering directly and those units entering indirectly in producing item i

$r_{ii} = 1$, by definition

Hence

$$r_{ij} = \begin{cases} \sum_{k=1}^{n} b_{ik} r_{kj} & \text{if } i \neq j \\ 1 & \text{if } i = j \end{cases}$$

Therefore,

$$R = BR + I$$
$$R = (I - B)^{-1} \tag{5.2}$$

In effect, the total requirements matrix can be found by inverting $(I - B)$. Since B is triangular, $(I - B)$ is a triangular matrix with determinant 1. Such a matrix is nonsingular and will always have an inverse; hence, R can always be found. We define

d = vector of final demand which includes the demand for end production, spare subassemblies, and spare components

x = total production requirements vector

Then

$$x = dR = d(I - B)^{-1} \qquad (5.3)$$

and for any product vector of forecasted demand, d, one can determine the total production vector of end products, subassemblies, and components, x.

EXAMPLE 5-2

A company has the product structure illustrated in Fig. 5-1. Demand for end products and spare assemblies is given in Table 5-4. Compute the total production vector.

Table 5-4.

Item	Units Demanded
1	20
2	30
A	0
D	10
B	0
C	5
γ	0
α	0
β	0

$$d = (20 \quad 30 \quad 0 \quad 10 \quad 0 \quad 5 \quad 0 \quad 0 \quad 0)$$

SOLUTION

Using Eq. (5.2), we determine the total requirements matrix.

$R = (I - B)^{-1} =$

	1	2	A	D	B	C	γ	α	β
1	1		2		2	5	2	9	15
2		1		1	3	7	6	7	21
A			1		1	2	1	4	6
D				1	2	5	2	5	15
B					1	2	1	2	6
C						1		1	3
γ							1		
α								1	
β									1

The total production vector is

$$x = d(I - B)^{-1} = (20 \quad 30 \quad 40 \quad 40 \quad 150 \quad 365 \quad 240 \quad 445 \quad 1095)$$

5.5 Computing Requirements Using Submatrix Structure

The BOM matrix of Table 5-1 has a clear submatrix structure as shown in Table 5-5.

Table 5-5. SUBMATRIX STRUCTURE OF B

	e	s	p
e	0	B^{es}	B^{ep}
$B = s$	0	B^{ss}	B^{sp}
p	0	0	0

where e = end product
s = subassemblies
p = parts

Matrix inversion is a tedious process and the inversion of large matrices usually leads to computational round-off errors. One can minimize the computational effort and errors by taking advantage of the submatrix structure of B. Since

$$(I - B) = \begin{bmatrix} I^{ee} & -B^{es} & -B^{ep} \\ 0 & I^{ss} - B^{ss} & -B^{sp} \\ 0 & 0 & I^{pp} \end{bmatrix}$$

The resulting requirements matrix will have the form:

$$R = (I - B)^{-1} = \begin{bmatrix} I^{ee} & R^{es} & R^{ep} \\ 0 & R^{ss} & R^{sp} \\ 0 & 0 & I^{pp} \end{bmatrix}$$

Values for the R submatrix elements can be solved by noting from Eq. (5.2).

$$R(I - B) = I$$

$$\begin{bmatrix} I^{ee} & R^{es} & R^{ep} \\ 0 & R^{ss} & R^{sp} \\ 0 & 0 & I^{pp} \end{bmatrix} \begin{bmatrix} I^{ee} & -B^{es} & -B^{ep} \\ 0 & I^{ss} - B^{ss} & -B^{sp} \\ 0 & 0 & I^{pp} \end{bmatrix} = \begin{bmatrix} I^{ee} & 0 & 0 \\ 0 & I^{ss} & 0 \\ 0 & 0 & I^{pp} \end{bmatrix}$$

We apply matrix multiplication to elements (2, 2), (1, 2), (2, 3), and (3, 1) as follows:

(2, 2):

$$R^{ss}(I^{ss} - B^{ss}) = I^{ss}$$
$$R^{ss} = (I^{ss} - B^{ss})^{-1}$$

(5.4)

(1, 2):

$$I^{ee}(-B^{es}) + R^{es}(I^{ss} - B^{ss}) = 0$$
$$R^{es} = B^{es}(I^{ss} - B^{ss})^{-1}$$
$$R^{es} = B^{es}R^{ss}$$

(5.5)

(2, 3):

$$R^{ss}(-B^{sp}) + R^{sp}I^{pp} = 0$$
$$R^{sp} = R^{ss}B^{sp}$$

(5.6)

(3, 1) transpose:

$$I^{ee}R^{ep} - B^{es}R^{sp} - B^{ep}I^{pp} = 0$$
$$R^{ep} = B^{ep} + B^{es}R^{sp}$$

(5.7)

The requirements matrix can now be found with matrix inversion limited only to the ss submatrix.

EXAMPLE 5-3

Use the submatrix structure of the BOM matrix in Table 5-1 to solve for the total requirements matrix R.

SOLUTION

$$R^{ss} = \begin{bmatrix} 1 & 0 & -1 & 0 \\ 0 & 1 & -2 & -1 \\ 0 & 0 & 1 & -2 \\ 0 & 0 & 0 & 1 \end{bmatrix}^{-1} = \begin{bmatrix} 1 & 0 & 1 & 2 \\ 0 & 1 & 2 & 5 \\ 0 & 0 & 1 & 2 \\ 0 & 0 & 0 & 1 \end{bmatrix}$$

$$R^{es} = \begin{bmatrix} 2 & 0 & 0 & 1 \\ 0 & 1 & 1 & 0 \end{bmatrix} \begin{bmatrix} 1 & 0 & 1 & 2 \\ 0 & 1 & 2 & 5 \\ 0 & 0 & 1 & 2 \\ 0 & 0 & 0 & 1 \end{bmatrix} = \begin{bmatrix} 2 & 0 & 2 & 5 \\ 0 & 1 & 3 & 7 \end{bmatrix}$$

$$R^{sp} = \begin{bmatrix} 1 & 0 & 1 & 2 \\ 0 & 1 & 2 & 5 \\ 0 & 0 & 1 & 2 \\ 0 & 0 & 0 & 1 \end{bmatrix} \begin{bmatrix} 0 & 2 & 0 \\ 0 & 0 & 0 \\ 1 & 0 & 0 \\ 0 & 1 & 3 \end{bmatrix} = \begin{bmatrix} 1 & 4 & 6 \\ 2 & 5 & 15 \\ 1 & 2 & 6 \\ 0 & 1 & 3 \end{bmatrix}$$

$$R^{ep} = \begin{bmatrix} 0 & 0 & 0 \\ 3 & 0 & 0 \end{bmatrix} + \begin{bmatrix} 2 & 0 & 0 & 1 \\ 0 & 1 & 1 & 0 \end{bmatrix} \begin{bmatrix} 1 & 4 & 6 \\ 2 & 5 & 15 \\ 1 & 2 & 6 \\ 0 & 1 & 3 \end{bmatrix} = \begin{bmatrix} 2 & 9 & 15 \\ 6 & 7 & 21 \end{bmatrix}$$

5.6 Engineering Changes

A problem in working with the BOM and requirements matrix is that product specifications are always changing. This creates a computational problem if the entire requirements matrix has to be regenerated each time a product change is made. Therefore, it is useful to find a way to update the requirements matrix with a minimum of burdensome computations.

Beginning with Eq. (5.2), $R = (I - B)^{-1}$, the postengineering change matrix, $R + \Delta R$, can be expressed as

$$R + \Delta R = (I - (B + \Delta B))^{-1} = (I - B - \Delta B)^{-1} \tag{5.8}$$

Multiplying Eqs. (5.2) and (5.8) through to eliminate inverses, we obtain

$$I = (I - B)R \tag{5.9}$$

$$I = (R + \Delta R)(I - B - \Delta B) \tag{5.10}$$

Multiplying Eq. (5.9) by $R + \Delta R$ and Eq. (5.10) by R, we obtain

$$(R + \Delta R)I = (R + \Delta R)(I - B)R \tag{5.11}$$

$$RI = (R + \Delta R)(I - B - \Delta B)R \tag{5.12}$$

Subtracting Eq. (5.12) from Eq. (5.11) yields the following result:

$$\Delta R = R \, \Delta B \, R + \Delta R \, \Delta B \, R \tag{5.13}$$

Equation (5.13) can be further simplified. We note that in the BOM matrix, an item is not a component of itself. If we change the composition of product i in the BOM matrix, the column in the ΔR matrix corresponding to item i will have zero entries. Hence, if we consider a single engineering change where ΔB has nonzero elements in only one column, we may write

$$\Delta R = R \, \Delta B \, R \tag{5.14}$$

and

$$R_{new} = R_{old} + \Delta R \tag{5.15}$$

EXAMPLE 5-4

Product 1 in Fig. 5-1 has been redesigned as follows:

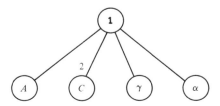

Compute the change in the requirements matrix.

SOLUTION

$$\Delta B =$$

	1	2	A	D	B	C	γ	α	β
1			−1			1	1	1	
2									
A									
D									
B									
C									
γ									
α									
β									

$$\Delta R = R \ \Delta B \ R =$$

	1	2	A	D	B	C	γ	α	β
1			−1		−1	−1		−2	−3
2									
A									
D									
B									
C									
γ									
α									
β									

5.7. Material Requirements Planning System

The bill of materials structure is part of the foundation of an MRP system. In addition to the bill of materials data base, an accurate data base of current inventory status, product routing and production lead time by manufactured part, and a master schedule are necessary in order to run this planning system. Figure 5-2 illustrates the minimum input requirements.

A preliminary step in operating the system is to determine the length of the planning time horizon. As a practical matter, the time horizon should not exceed the ability to forecast and it is preferable to use a time horizon in which existing orders are relatively firm. One must also break down the horizon period into time intervals, called *time buckets*. These planning periods are typically weekly, biweekly, or monthly, depending on the lead time between departments and the degree of control desired.

The MRP planning process is most easily described through illustration. For this purpose, we will use the product structure previously given in Fig. 5-1 and the additional data set forth in Tables 5-6 and 5-7. Table 5-6 gives the

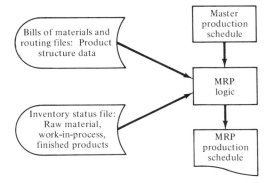

Figure 5-2. Elements of an MRP planning system.

current inventory status and the necessary reorder lead-time information. Table 5-7 is the master production schedule of items independently demanded over a 9-week planning horizon.

Table 5-8 illustrates the level 0 portion of a typical MRP report. The *gross requirements* for an item during a time period is defined as the total amount of that item that will have to be disbursed during that period. For end items 1 and 2, the gross requirements are given by the first two rows of Table 5-7.

MRP systems are run on a rolling horizon basis with periodic updates. For any particular computer run that updates the schedule, there will usually be an outstanding order for lower-level subassemblies and components that was placed during a previous schedule run. This scheduled delivery of lower-level purchased and manufactured parts is called a *scheduled receipt*. A scheduled receipt of 1 unit at a given level includes all the lower-level subassemblies and parts required to complete 1 unit of the item at that level.

Table 5-6. CURRENT INVENTORY STATUS AND PRODUCTION LEAD TIMES

Level	Item	On-Hand Inventory at t = 0	Reorder Lead Time (weeks)
0	1	120	1
0	2	85	1
1	A	0	2
1	D	10	2
2	B	500	1
3	C	160	1
3	γ	0	2
4	α	1200	1
4	β	4000	2

Table 5-7. PLANNING HORIZON MASTER SCHEDULE IN WEEKLY TIME BUCKETS (UNITS OF ITEMS TO BE PRODUCED)

Item	Period								
	1	2	3	4	5	6	7	8	9
1	50	20	30	40	40	30	25	15	30
2	20	30	25	35	10	35	20	25	30
A								15	
D		10		10					
B					20				100
C		5							
γ									
α									
β									

The *expected inventory* is the beginning of the period inventory level. This is computed as follows: Let

I_{jt} = expected on-hand inventory of item j at the beginning of period t (item backlogging is not allowed)

S_{jt} = scheduled receipt of item j during period t

G_{jt} = gross requirement of item j during period t

Then the expected inventory at the beginning of period t is

$$I_{jt} = \max \{0, I_{j,t-1} + S_{j,t-1} - G_{j,t-1}\} \qquad (5.16)$$

EXAMPLE 5-5

What is the expected inventory for item 1 in period 5?

Table 5-8. GROSS AND NET REQUIREMENTS REPORT, LEVEL 0 ACTIVITY

	Period								
	1	2	3	4	5	6	7	8	9
Item 1, Level 0									
Gross requirements	50	20	30	40	40	30	25	15	30
Scheduled receipts				120					
Expected inventory	120	70	50	20	100	60	30	5	0
Net requirements								10	30
Planned order release			120				120		
Item 2, Level 0									
Gross requirements	20	30	25	35	10	35	20	25	30
Scheduled receipts				100					
Expected inventory	85	65	35	10	75	65	30	10	0
Net requirements								15	30
Planned order release			100				100		

SOLUTION

$$I_{1,5} = I_{1,4} + S_{1,4} - G_{1,4}$$
$$= 20 + 120 - 40$$
$$= 100$$

In any period, the net requirements are the items needed to meet gross requirements that will not be available from on-hand inventory or scheduled receipts. The existence of a net requirement signals the potential for a backorder situation; hence, production control plans to initiate the activities required for a *planned order release* in time to prevent the stockout from occurring. The net requirement is computed as follows: Let

N_{jt} = net requirements for item j in period t; then
$N_{jt} = \max \{0, G_{jt} - S_{jt} - I_{jt}\}$

EXAMPLE 5-6

Compute the net requirements for item 1 in period 8.

SOLUTION

$$N_{1,8} = G_{1,8} - S_{1,8} - I_{1,8}$$
$$= 15 - 0 - 5$$
$$= 10$$

The planned order release for item 1 is required in period 7 because there is a 1-week lead time to obtain delivery (see Table 5-6). The planned order release quantity of 120 is based on lot sizing rules for item 1. There are many schemes in use for MRP lot sizing; they will be discussed in Section 5.8.

Production activity at lower levels of the product structure is based primarily on the planned order releases at level 0. The planned order release of 120 units of item 1 and 100 units of item 2 in week 7 create the gross requirements in week 7 for subassemblies and components used directly in the construction of items 1 and 2. These are subassemblies A, D, B, and C and component γ. These requirements can be computed using the method of Section 5.3.

EXAMPLE 5-7

Compute the week 7 gross requirements for A, D, B, C, and γ.

SOLUTION

Using Eq. (5.1), we compute the gross production requirements generated by level 0 as follows:

$$dd(0) = d_0 \times B$$

$$dd(0) = (120 \quad 100 \quad 0 \quad 0 \quad 0 \quad 0 \quad 0 \quad 0 \quad 0) \begin{bmatrix} 0 & 0 & 2 & 0 & 0 & 1 & 0 & 0 & 0 \\ 0 & 0 & 0 & 1 & 1 & 0 & 3 & 0 & 0 \\ 0 & 0 & 0 & 0 & 1 & 0 & 0 & 2 & 0 \\ 0 & 0 & 0 & 0 & 2 & 1 & 0 & 0 & 0 \\ 0 & 0 & 0 & 0 & 0 & 2 & 1 & 0 & 0 \\ 0 & 0 & 0 & 0 & 0 & 0 & 0 & 1 & 3 \\ 0 & 0 & 0 & 0 & 0 & 0 & 0 & 0 & 0 \\ 0 & 0 & 0 & 0 & 0 & 0 & 0 & 0 & 0 \\ 0 & 0 & 0 & 0 & 0 & 0 & 0 & 0 & 0 \end{bmatrix}$$

$$x = (0 \quad 0 \quad 240 \quad 100 \quad 100 \quad 120 \quad 300 \quad 0 \quad 0)$$

WEEK 7 REQUIREMENTS

Item	From Level 0	Other	Total
A	240	0	240
D	100	0	100
B	100	0	100
C	120	0	120
γ	300	0	300

The solution to Example 5-7 is found under week 7 in Table 5-9, which displays a partial gross and net requirements report. Gross requirements arising from independent demand for subassemblies and components (see Table 5-7) have been added.

The level 1 planned order releases have been computed using the procedure previously explained for level 0. A 2-week lead time applies at level 1. Net requirements at level 2 cannot be computed until the gross requirements due to planned order releases at level 1 have been computed.

EXAMPLE 5-8

From Table 5-9, compute the week 5 lower-level gross requirements implied by the week 5 planned order releases at level 1.

SOLUTION

$$dd(1) = d_1 \times B$$

$$dd(1) = (0 \quad 0 \quad 240 \quad 100 \quad 0 \quad 0 \quad 0 \quad 0 \quad 0) \times B$$

$$x = (0 \quad 0 \quad 0 \quad 0 \quad 440 \quad 100 \quad 0 \quad 480 \quad 0)$$

WEEK 5 REQUIREMENTS

Item	From Level 1	Other	Total
B	440	20	460
C	100	0	100
α	480	0	480

Table 5-9. PARTIAL GROSS AND NET REQUIREMENTS REPORT

	Period								
	1	*2*	*3*	*4*	*5*	*6*	*7*	*8*	*9*
Item 1, Level 0									
Gross requirements	50	20	30	40	40	30	25	15	30
Scheduled receipts				120					
Expected inventory	120	70	50	20	100	60	30	5	0
Net requirements								10	30
Planned order release			120				120		
Item 2, Level 0									
Gross requirements	20	30	25	35	10	35	20	25	30
Scheduled receipts				100					
Expected inventory	85	65	35	10	75	65	30	10	0
Net requirements								15	30
Planned order release			100				100		
Item A, Level 1									
Gross requirements	0	0	240	0	0	0	240	15	0
Scheduled receipts			240						
Expected inventory	0	0	0	0	0	0	0	0	
Net requirements							240	15	
Planned order release	240				240	15			
Item D, Level 1									
Gross requirements	0	10	100	10	0	0	100	0	0
Scheduled receipts		10	100	10					
Expected inventory	0	0	0	0	0	0	0	0	0
Net requirements							100		
Planned order release	100	10			100				
Item B, Level 2									
Gross requirements			100		20		100		100
Scheduled receipts									
Expected inventory									
Net requirements									
Planned order release									
Item C, Level 3									
Gross requirements		5	120				120		
Scheduled receipts									
Expected inventory									
Net requirements									
Planned order release									
Item γ, Level 3									
Gross requirements			300				300		
Scheduled receipts									
Expected inventory									
Net requirements									
Planned order release									
Item α, Level 4									
Gross requirements									
Scheduled receipts									
Expected inventory									
Net requirements									
Planned order release									
Item β, Level 4									
Gross requirements									
Scheduled receipts									
Expected inventory									
Net requirements									
Planned order release									

Table 5-10. PARTIAL GROSS AND NET REQUIREMENTS REPORT

	Period								
	1	2	3	4	5	6	7	8	9
Item 1, Level 0									
Gross requirements	50	20	30	40	40	30	25	15	30
Scheduled receipts				120					
Expected inventory	120	70	50	20	100	60	30	5	0
Net requirements								10	30
Planned order release			120				120		
Item 2, Level 0									
Gross requirements	20	30	25	35	10	35	20	25	30
Scheduled receipts				100					
Expected inventory	85	65	35	10	75	65	30	10	0
Net Requirements								15	30
Planned order release			100				100		
Item A, Level 1									
Gross requirements	0	0	240	0	0	0	240	15	0
Scheduled receipts			240						
Expected inventory	0	0	0	0	0	0	0	0	
Net requirements							240	15	
Planned order release	240				240	15			
Item D, Level 1									
Gross requirements	0	10	100	10	0	0	100	0	0
Scheduled receipts		10	100	10					
Expected inventory	0	0	0	0	0	0	0	0	0
Net requirements							100		
Planned order release	100	10			100				
Item B, Level 2									
Gross requirements	440	20	100	0	460	15	100	0	100
Scheduled receipts	560								
Expected inventory	0	120	100	0	0	0	0	0	0
Net requirements					460	15	100		100
Planned order release				575				100	
Item C, Level 3									
Gross requirements	100	15	120		100		120		
Scheduled receipts									
Expected inventory									
Net requirements									
Planned order release									
Item γ, Level 3									
Gross requirements			300				300		
Scheduled receipts									
Expected inventory									
Net requirements									
Planned order release									
Item α, Level 4									
Gross requirements	480				480	30			
Scheduled receipts									
Expected inventory									
Net requirements									
Planned order release									
Item β, Level 4									
Gross requirements									
Scheduled receipts									
Expected inventory									
Net requirements									
Planned order release									

Table 5-11. COMPLETED GROSS AND NET REQUIREMENTS REPORT

	Period								
	1	2	3	4	5	6	7	8	9
Item 1, Level 0									
Gross requirements	50	20	30	40	40	30	25	15	30
Scheduled receipts				120					
Expected inventory	120	70	50	20	100	60	30	5	0
Net requirements								10	30
Planned order release			120				120		
Item 2, Level 0									
Gross requirements	20	30	25	35	10	35	20	25	30
Scheduled receipts				100					
Expected inventory	85	65	35	10	75	65	30	10	0
Net requirements								15	30
Planned order release			100				100		
Item A, Level 1									
Gross requirements	0	0	240	0	0	0	240	15	0
Scheduled receipts			240						
Expected inventory	0	0	0	0	0	0	0	0	0
Net requirements							240	15	
Planned order release	240				240	15			
Item D, Level 1									
Gross requirements	0	10	100	10	0	0	100	0	0
Scheduled receipts		10	100	10					
Expected inventory	0	0	0	0	0	0	0	0	0
Net requirements							100		
Planned order release	100	10			100				
Item B, Level 2									
Gross requirements	440	20	100	0	460	15	100	0	100
Scheduled receipts	560								
Expected inventory	0	120	100	0	0	0	0	0	0
Net requirements					460	15	100		100
Planned order release				575				100	
Item C, Level 3									
Gross requirements	100	15	120	1150	100	0	120	200	0
Scheduled receipts	115		120						
Expected inventory	0	15	0	0	0	0	0	0	0
Net requirements					1150	100		120	200
Planned order release		120	1250				320		
Item γ, Level 3									
Gross requirements	0	0	300	575	0	0	300	100	0
Scheduled receipts			300						
Expected inventory	0	0	0	0	0	0	0	0	0
Net requirements				575			300	100	
Planned order release	300	575			300	100			
Item α, Level 4									
Gross requirements	480	120	1250	0	480	350	0	0	0
Scheduled receipts									
Expected inventory	3000	2520	2400	1150	1150	670	320	320	320
Net requirements									
Planned order release									
Item β, Level 4									
Gross requirement	0	360	3750	0	0	960	0	0	0
Scheduled receipts									
Expected inventory	4000	4000	3640	0	0	0	0	0	0
Net requirements			110			960			
Planned order release	5000								

Table 5-10 shows the next stage partial completion of the gross and net requirements report, with net requirements and planned order arrival computations completed through level 2. This process is continued iteratively through all levels and for each time period. Table 5-11 is the complete gross and net requirements report, which was generated by successive iterations of the procedure described.

5.8. Lot Sizing

There are several methods employed for lot sizing in MRP systems. We will describe a few of them in the paragraphs that follow.

Lot-for-lot is the name given to the method that orders exactly what is required in each period. For example, in Table 5-11, a lot-for-lot ordering policy was used for subassemblies *A* and *D* at level 1. Planned order releases were matched to net requirements. Lot-for-lot ordering implicitly assumes that the setup or ordering cost is much smaller than the carrying cost.

Deterministic EOQ models are sometimes employed to determine the size of order quantities. The reader will recall that EOQ models assumed demand is constant, which is usually not the case. Where the variance to mean ratio of demand is very small, the EOQ model may give reasonable answers and it has the advantage of taking setup cost into consideration.

There are two marginal cost methods often used in MRP: (1) *least unit cost* and (2) *least total cost*. To illustrate the nature of the approach, we will assume the least-unit-cost method.

Least-unit-cost computations begin by placing an order for the first net requirement. Then the question is asked: Should the size of the order be extended to cover the next period's net requirements? The answer is "yes" only if the unit cost for so doing is lower than just ordering for the first period.

Consider the net requirements for Item *C* in periods 4 through 8 in Table 5-11. These have been reproduced in column 2 of Table 5-12. Assuming that the only period of production was period 4 (assumed for iteration 1), column 3 shows the prospective lot size appropriate to ordering a quantity to cover net requirements through each prospective period. Thus, if a quantity of Item *C* were to be ordered to cover net requirements through period 5, the order size would be $1150 + 100 = 1250$.

Column 4 shows the number of periods that the net requirements of a period would be held in inventory if it were part of the order which included the net requirements of period 4. Hence, if the order size were 1250, 100 units would be carried for 1 period.

Carrying cost (column 5) is computed as the sum of the unit periods being carried times the holding cost per unit period; also, column 6 = (column

Table 5-12. ILLUSTRATION OF LEAST-UNIT-COST LOT-SIZING COMPUTATIONS FOR ITEM C*

(1) Period	(2) Net Requirements	(3) Prospective Lot Size	(4) Carried in Inventory	(5) Carrying Cost per Lot	(6) Dollars per Unit	(7) Setup Unit Cost	(8) Total Unit Cost
Iteration 1							
4	1150	1150	0	0	0	0.87	0.87
5	100	1250	1	60	0.05	0.80	0.85
6	0	—	2	—	—	—	—
7	120	1370	3	276	0.20	0.73	0.93
8	200	1570	4	756	0.48	0.64	1.12
Iteration 2							
7	120	120	0	0	0	8.33	8.33
8	200	320	1	120	0.38	3.13	3.51

*Setup cost = $1000
Holding cost = $0.60/unit-period
Unit cost = $1

5)/(column 3). The unit setup cost (column 7) is the (total setup cost)/(column 3), while column 8 = column 6 plus column 7.

The average unit cost has a minimum at $0.85, which means the order release for periods 4 and 5 will be combined. The computational search for a minimum begins again in period 7; this is shown in iteration 2.

Marginal-cost computations cannot be relied on to give a global optimal solution over the production planning horizon. The reason for this is that subsequent lot-size decisions are made subject to earlier decisions. If an earlier decision had been different, it might have led to better later decisions. Least-unit-cost searches for a series of local optima but does not guarantee a global optima.

From a theoretical point of view, the preferred lot-sizing technique for obtaining global optima in MRP is the *Wagner–Whitin algorithm*. The Wagner–Whitin algorithm, described in Chapter 4, overcomes the deficiency in marginal-cost techniques by employing a dynamic programming recursion that evaluates all admissible alternative production schedules.

EXAMPLE 5-9

Apply the Wagner–Whitin algorithm to lot sizing for item B in Table 5-11. The relevant costs are:

$$\text{setup cost} = \$150$$

$$\text{holding cost} = \$0.60/\text{unit-period}$$

$$\text{unit cost} = \$5$$

SOLUTION

Period	Index t	N_t	h_t	c_t
5	1	460	0.60	5
6	2	15	0.60	5
7	3	100	0.60	5
8	4	0	0.60	5
9	5	100	0.60	5

Applying the Wagner–Whitin recursion, we obtain

$$Z_1 = Z_0 + C_{01} = A_1 + c_1 N_1 = 150 + 5(460) = 2450^*$$

$$Z_2 = \begin{cases} Z_0 + C_{02} = A_1 + c_1(N_1 + N_2) + h_1 N_2 \\ \qquad = 150 + 5(475) + (0.6)(15) = 2534^* \\ Z_1 + C_{12} = Z_1 + A_2 + c_2 N_2 \\ \qquad = 2450 + 150 + 5(15) = 2675 \end{cases}$$

$$Z_3 = \begin{cases} Z_0 + C_{03} = A_1 + c_1(N_1 + N_2 + N_3) + h_1(N_2 + N_3) + h_2 N_3 \\ \qquad = 150 + 5(575) + (0.6)(115) + (0.6)(100) = 3154^* \\ Z_1 + C_{13} = Z_1 + A_2 + c_2(N_2 + N_3) + h_2 N_3 \\ \qquad = 2450 + 150 + 5(115) + (0.6)(100) = 3235 \\ Z_2 + C_{23} = Z_2 + A_3 + c_3 N_3 \\ \qquad = 2534 + 150 + 5(100) = 3184 \end{cases}$$

$$Z_5 = \begin{cases} Z_0 + C_{05} = A_1 + c_1(N_1 + N_2 + N_3 + N_5) + h_1(N_2 \\ \qquad + N_3 + N_5) + h_2(N_3 + N_5) + h_3 N_5 + h_4 N_5 \\ \qquad = 150 + 5(675) + (0.6)(215) + (0.6)(200) \\ \qquad + (0.6)(100) + (0.6)(100) = 3894 \\ Z_1 + C_{15} = Z_1 + A_2 + c_2(N_2 + N_3 + N_5) \\ \qquad + h_2(N_3 + N_5) + h_3 N_5 + h_4 N_5 \\ \qquad = 2450 + 150 + 5(215) + (0.6)(200) + (0.6)(100) + (0.6)(100) \\ \qquad = 3915 \\ Z_2 + C_{25} = Z_2 + A_3 + c_3(N_3 + N_5) \\ \qquad + h_3 N_5 + h_4 N_5 \\ \qquad = 2534 + 150 + 5(200) + (0.6)(100) + (0.6)(100) = 3804^* \\ Z_3 + C_{35} = Z_3 + A_4 + c_4 N_5 \\ \qquad = 3154 + 150 + 5(100) = 3804^* \end{cases}$$

Recall from Chapter 4 the usual practice of tracing the DP solution backward from the last stage; then there are two optimal solutions, each with a total cost of $3804. Either

$$Q_1 = 575, \quad Q_2 = 0, \quad Q_3 = 0, \quad Q_4 = 0, \quad Q_5 = 100$$

which is the ordering policy used in Table 5-11, or

$$Q_1 = 475, \quad Q_2 = 0, \quad Q_3 = 200, \quad Q_4 = 0, \quad Q_5 = 0$$

5.9. Software Structure of MRP

There are many commercially available MRP software packages which provide the results shown in this chapter as well as other informative reports. Figure 5-3 illustrates the general structure of these planning systems.

As stated previously, the major components of the system are the master production schedule, the bill of materials and routing files, the inventory status file, and the MRP software that provides the logic and performs the computations. The bill of materials and routing files are updated as required through an engineering change module and inventory status is kept current by posting inventory transactions daily.

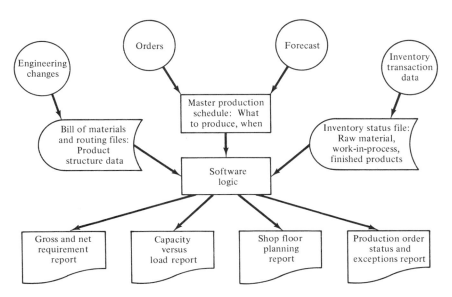

Figure 5-3. Typical structure of commercial MRP-based planning system.

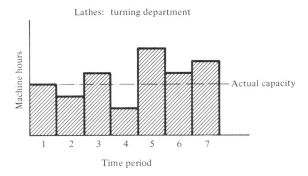

Figure 5-4. Capacity versus load report, lathes.

The gross and net requirements report illustrated in this chapter is only one of a number of reports generated in a typical MRP scheduling system. Although the gross and net requirements report indicates the timing of order releases required to meet the master schedule, it does not determine whether the resources required by the work centers are available to produce these orders. This is done in the *capacity versus load report.*

Using the schedule of component and subassembly manufacture in the gross and net requirements report and the standard times and routings for those items in the routing file, one can compute the period-by-period load for the various work centers in the manufacturing plant. Figure 5-4 is a generalized illustration of a capacity planning aid for the lathe department. If the production of components and subassemblies is performed as scheduled in the gross and net requirements report, the load of required lathe hours versus actual capacity will be as shown in Fig. 5-4. There are periods of under-capacity utilization (periods 2 and 4) and periods of overcapacity utilization (periods 1, 3, 5, 6, and 7). In general, there appears to be insufficient lathe capacity to carry out the given plan in-plant. The purpose of this report is to alert production management as to the infeasibility of the given plan. Typical actions that may be taken at this point are to plan overtime in the lathe department in order to increase available capacity, subcontract some of the machining to an outside vendor in order to reduce the load, or revise the master schedule and regenerate a new plan. It is the purpose of the capacity planning report to check on overall feasibility of the plan over the planning horizon by time period.

A more detailed operational report provided by most commercial software is the *shop floor planning report*, illustrated in Table 5-13. The purpose of this report is to provide shop floor supervision with a daily posting of jobs queued at their work centers. It is typical to sort and list jobs by due date, where the due date indicates when machining at that department should be

Table 5-13. SHOP FLOOR PLANNING REPORT

Cost center: 963, machine shop
Machine: 41, lathe, engine (18 to 34 in.)

Part Number	Operation	Hours	Production Due Date
622496-4	Turn 2-in. round stock	6	100
738242-5	Recess shaft $\frac{5}{8}$ in.	12	102
733921-4	Bore ID $\frac{1}{4}$ in.	10	103

completed in order to meet the schedule indicated by the gross and net requirements report. The shop floor supervisor will then release jobs to machines, taking into consideration the hours of machining required and the due dates for the items.

In addition to the reports illustrated above, a typical commercial system provides additional information under the general heading of *exceptions reports*. For example, jobs that should be expedited or deexpedited are highlighted. The production of a component should be deexpedited when delays have occurred in the production of another component and both of these components are simultaneously required for the same subassembly at a higher level. Deexpediting holds down work-in-process inventory.

Finally, it should be pointed out that commercial software systems process information using data base structures. Files are organized around a sequence of keys (i.e., fields that identify that file). Files that are related are linked to each other by a method called *chaining*. For example, a level 4 component file that is used in several higher-level items would be linked to those files via a chain structure. This eliminates the necessity of carrying duplicate files for subassemblies and components used in more than one end product. For a detailed description of data base technology applied to MRP data processing, see Ref. 12.

Matrix representation and manipulation of the BOM is a concise tool to illustrate the structure of the relationships of which MRP takes advantage. However, MRP software uses *list processing* techniques as opposed to matrix algebra to assemble gross and net requirements by level. This method is faster and less error prone than matrix inversion.

The main benefit of the MRP approach to discrete parts production planning is in its ability to provide feasible schedules for a very complex production system. MRP is not an optimization technique; MRP simply tries to schedule all the activities required to meet a given master schedule, while holding down work-in-process inventory. If infeasibility occurs, production management must produce a new master schedule and generate another plan or find alternative sources of production capacity.

PROBLEMS

5-1. A manufacturer is producing cranes for a mini-load automated storage and retrieval system. The crane assembly is shown in Fig. 5-5 and the indented bill of materials structure is given below.

PRODUCT: STORAGE/RETRIEVAL CRANE

Level	Description	Quantity
A	Crane frame assembly	1
A1	Frame assembly	1
A11	Cross bracing	8
A12	Crane frame angles	4
A13	Top plate	1
A14	Bottom plate	1
A15	Bearings, 1.375 outer diameter	1
A16	Picking head guide	8
A2	Motor assembly	1
A21	Stepping motor, 750 ounces/inch	1
A22	Motor supports	2
A23	Sprocket	2
A24	Chain	2
A25	Sprocket support	4
B	Order picking head subassembly	1
B1	Bucket frame	1
B11	Picking head frame angles	8
B12	Chain support	1
B2	Motor assembly	1
B22	Moving plate	1
B23	Fixed plate	1
B24	Solenoid	2
B25	Picking lever	2
B26	Spring	2
B27	Rack and Pinion	2
B28	Stepping motor, 120 ounces/inch	1

Required:
   1. Construct the BOM matrix.
   2. Compute the total requirements matrix.

5-2. Consider the following product structures:

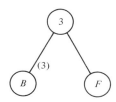

Subassemblies:

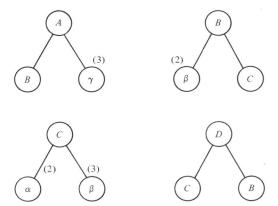

Items have the following six-period demand schedule:

DEMAND BY ITEM

	Period					
Item	*1*	*2*	*3*	*4*	*5*	*6*
1	40	20	10	60	10	30
2	20	10	30	50	70	40
A						
B				10		
C						10
D						
α					20	
β						
γ						

On-hand inventory and production lead times are as follows:

Item	*On-hand Inventory*	*Lead Time (periods)*
1	80	1
2	40	1
A	30	1
B	100	1
C	100	1
D	100	1
α	200	1
β	300	1
γ	500	1

Develop the gross and net requirements report for the 6-week period. Assume lot-for-lot lot sizing.

5-6. For the situation given in Prob. 5-5, assume the following production worker requirements per item produced:

Item	Worker-Hours Required
1	2
2	4
A	1
B	1
C	2
D	2
α	3
β	1
γ	6

Assuming there are 1550 worker-hours available per time period, check the overall capacity constraint against your production plan and modify it accordingly.

REFERENCES

[1] ANDERSON, JOHN C., ROGER S. SCHROEDER, SHARON E. TUPY, AND EDNA M. WHITE, "Material Requirements Planning Systems: The State of the Art," *Production and Inventory Management*, fourth Quarter, 1982.

[2] ANDERSSON, HAKAN, SVEN AXSATER, AND HENRIK JONSSON, "Hierarchical Material Requirements Planning," *International Journal of Production Research*, Vol. 19, No. 1, Jan./Feb. 1981.

[3] BERRY, WILLIAM L., "Lot Sizing Procedures for Requirements Planning Systems: A Framework for Analysis," *Production and Inventory Management*, second Quarter, 1972.

[4] BEVIS, GEORGE E., "A Management Viewpoint on the Implementation of a MRP System," *Production and Inventory Management*, first Quarter, 1976.

[5] BÜCHEL, ALFRED, "Stochastic Material Requirements Planning for Optional Parts," *International Journal of Production Research*, Vol. 21, No. 4, July/Aug. 1983.

[6] ELMAGHRABY, SALAH E., "A Note on the Explosion and Netting Problem in the Planning of Materials Requirements," *Operations Research*, Vol. 11, no. 4 (July-Aug. 1963).

[7] GIFFLER, BERNARD, "Mathematical Solutions of Parts Requirements Problems," *Management Science*, 11, no. 9 (July) 1965.

[8] GLEIBERMAN, LEON, "Engineering Change of the Total Requirements Matrix for a Bill of Materials," *Management Science*, Vol. 10, no. 3, (April 1964).

[9] MILLER, JEFFREY, G., "Behind the Growth in Materials Requirements Planning," *Harvard Business Review*, (Sept.–Oct. 1975).

[10] MORE, STEVEN M., "MRP and Least Total Cost Method of Lot-Sizing," *Production and Inventory Management*, second Quarter, 1974.

[11] "MRP II Software—the Package You Need", *Modern Materials Handling*, 37, no. 17, (November), 1982.

[12] ORLICKY, JOSEPH, A., *Material Requirement Planning*, Highstown, N.J.: McGraw-Hill Book Co., 1975.

[13] ORLICKY, JOSEPH A., GEORGE W. PLOSSL, AND OLIVER W. WRIGHT, "Structuring the Bill of Material for MRP," *Production and Inventory Management*, fourth Quarter, 1972.

[14] VAZSONYI, ANDREW, "A Mathematical Model of Production Scheduling", *Scientific Programming in Business and Industry*, chapter 13, New York, N.Y.: John Wiley & Sons, 1958.

[15] VAZSONYI, ANDREW, "The Use of Mathematics in Production and Inventory Control," *Management Science*, 1, nos. 1, 3, and 4, 1955.

[16] WAGNER, HARVEY M., "Dynamic Version of the Economic Lot Size Model," *Management Science*, Vol. 5, no. 1 (Sept. 1958).

[17] YELLE, LOUIS E., "Materials Requirements Lot Sizing: A Multi-level Approach," *International Journal of Production Research*, Vol. 17, No. 3, May/June 1979.

PROJECT PLANNING
AND SCHEDULING

6.1. Introduction

A project is defined as a combination of interrelated activities that must be executed in a particular order to complete an entire task. In the past, the planning of large projects was considered a complex and difficult task. Up to the end of the nineteenth century, decision making was primarily dependent on the capabilities and experience of the managers.

The Gantt chart was the first scientific technique for project planning and scheduling. With the development of computers, new scheduling techniques came into existence. In the late 1950s, the network techniques of CPM (critical path method) and PERT (program evaluation and review technique) were developed concurrently. These two techniques are widely accepted and their use is well established in both government and industry. CPM and PERT are based on the assumption that there is an infinite number of resources available for the project's use at any time. This restricts their use in many practical situations.

In this chapter we present techniques for project planning and scheduling. These techniques are then utilized under two conditions: (1) when unlimited resources are available for the project activities, and (2) when resources are limited. We first present the techniques for project planning and scheduling with unlimited resources.

6

6.2. Project Planning and Scheduling: Unlimited Resources

Project management involves planning, organizing, staffing, controlling, monitoring, and directing a project to its successful completion. These activities can be achieved optimally by using CPM and PERT.

6.2.1. CPM: Background

The *critical path method* (CPM) was developed by Kelley and Walker [11]. It was used in the construction industry, where previous experience was the basis for time and cost estimates of the different phases of the project.

The basic feature of CPM is that it considers duration estimates over a range of cost levels and, as a result, provides a range of project durations with an associated range of project costs. Computational procedures for CPM can be used to establish the minimum cost for completing the project. In brief, CPM could be described mathematically as a deterministic, digraphic (directed graph), longest path network model.

6.2.2. PERT: Background

The *program evaluation and review technique* (PERT) was developed in 1959 as a joint effort of Booz, Allen and Hamilton and the U.S. Navy's Special Project Office as a research and development tool for the Polaris Missile project. With its use, this important project was completed approximately 24 months ahead of the originally scheduled completion date. The basic feature of PERT is the probabilistic estimation of activity times. A computational procedure then calculates an estimated overall project duration and derives a measure of certainty of meeting this estimate. PERT could be described mathematically as a probabilistic, digraphic (directed graph), longest path network model.

6.2.3. THE USE OF CPM/PERT IN INDUSTRY

In a survey conducted by Davis [5] among the top 400 construction firms in the United States, 80% of the respondent firms used critical path methods; however, not all users of the project networks were satisfied with the results. Davis noted that 16% of the construction firm respondents in the survey reported that they had been unsuccessful in achieving the various advantages attributed to CPM compared to 61% who reported "moderate success" and 15% who said they had been "highly successful."

In another survey, Gaither [8] found that PERT and CPM ranked above all other quantitative decision-making tools in terms of the percentage of firms who used them. In sampling 500 companies located in seven south-central states, Gaither found about half of them using one or more operations research techniques; among those which did, 69% reported using PERT with 67% using CPM. The next most used tools were linear programming, 57%; statistical analyses, 56%; and computer simulation, 52%.

6.2.4. CONSTRUCTION OF PROJECT NETWORKS

Any project network consists of two basic elements, *activities* and *events*, which are controlled by a *precedence* relationship. Before analyzing the project network, the following definitions and notations are introduced:

—*Activity:* a time-consuming effort required to complete a necessary segment of the project network. All activities must begin and end with an event.

—*Event:* a point of time signaling the beginning or completion of one or more activities. The event could also be defined as the point at which the successful completion of all preceding activities enables the dependent succeeding activities to start.

—*Precedence:* a term that describes the relationship between two or more activities in the network. For example, if an activity A precedes another activity B, activity A must be completed before activity B can start. This precedence relationship is usually written as $A < B$. It should be noted that the precedence is a transitive relationship; that is, if $A < B$, and $B < C$, then $A < C$.

Event i (i) $\xrightarrow{\text{Arc} \equiv \text{activity}}$ (j) Event j

Figure 6-1. A-on-A representation.

Placing an activity in the network is dependent on:

1. The precedence relationship between this activity and other activities
2. Activities that could be done coincident with this activity
3. Activities that could start upon the completion of this activity

There are two modes of representation of activities in a project network: the *activity-on-arc* representation and the *activity-on-node* representation. These representations are now explained.

Activity-on-arc representation (A-on-A). In this representation, activities are represented by arcs in the project network. An arrow is placed on the arc where the event (node) at the tail of the arrow represents the start of the activity and the event (node) at the head represents the termination (completion) of the activity. The description of an activity is written along the arc that represents this activity. Figure 6-1 shows a typical representation of an activity and events.

As indicated earlier, each activity begins and ends with an event (node) which could be graphically represented as a circle, square, or other geometric figures; and numbers, for identification, are written in the node. The following characteristics must be taken into consideration when constructing the graphical representation of the project network:

1. The nodes (events) are numbered such that an arrow leads from an event with a small number to an event with a larger one.
2. Each node must have at least one arc leading into and one arc leaving it, with the exception of the first node (origin or source) and last node (terminal or sink).

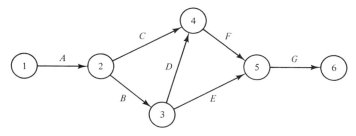

Figure 6-2. Example of a complete A-on-A network.

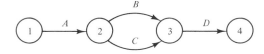

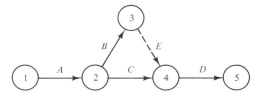

Figure 6-3. Use of a dummy activity in the network.

3. Cycles are not permitted; that is, an arc cannot lead into and leave the same node. Figure 6-2 shows an example of a complete activity network.
4. Any two events (nodes) may be connected by at most one arc. However, if two activities begin with the same event (node) and will lead into the same node, a dummy constraint is used to identify one of the activities. A *dummy constraint* is an arc on the network that shows a precedence relationship between activities and does not require a time-consuming activity to satisfy. Insertion of the dummy constraint prevents two or more activities from being identified by the same set of numbers. Also, dummy constraints can be used to eliminate a dependency relationship between activities that do not actually exist. Figure 6-3 illustrates the use of a dummy constraint (*E* is a dummy activity with a zero-time duration.) Throughout this text a dashed activity line indicates a dummy activity.

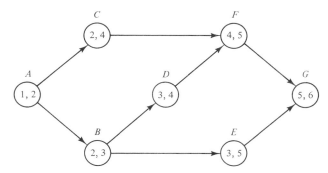

Figure 6-4. A-on-N.

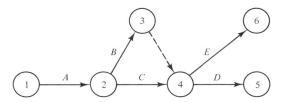

Figure 6-5. Incorrect representation of Example 6-1. *A* should not precede *B*; *C* should not precede *E*.

Activity-on-node representation (A-on-N). In this representation the activities are placed in the nodes, while the precedence relationship is represented by an arc in which the direction of the arrow specifies the precedence. In other words, *A-on-N* representation is the *dual* of the *A-on-A* representation. This representation mode is not as common as the *A-on-A* representation. The *A-on-A* representation will be used throughout this text. Figure 6-4 is the *A-on-N* representation of Fig. 6-2.

6.2.5. COMMON ERRORS IN NETWORK CONSTRUCTION

In this section we present some examples of common errors in constructing a network as well as the incorrect use of the dummy constraints.

EXAMPLE 6-1

Draw a project network with the following precedence relationships:

$$A < C \qquad B < D, E \qquad C < D$$

SOLUTION

Figures 6-5 and 6-6 represent an incorrect and correct representation of the network.

EXAMPLE 6-2

Draw a project network with the following precedence relationships:

$$A < C, F \qquad B < C, F \qquad D < E, F$$

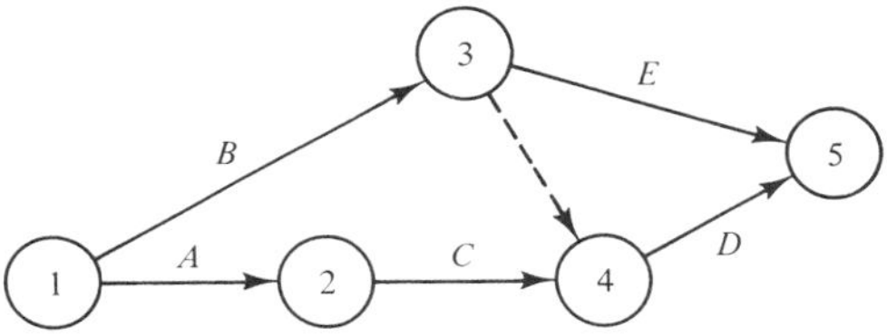

Figure 6-6. Correct presentation of Example 6-1.

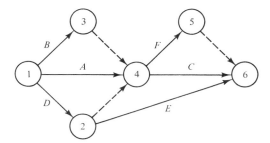

Figure 6-7. Incorrect representation of Example 6-2. *D* should not precede *C*.

SOLUTION

Figures 6-7 and 6-8 represent an incorrect and correct representation of the network.

6.2.6. CHECKING THE CONSISTENCY OF PRECEDENCE RELATIONSHIPS

After constructing the project network, one needs to check the consistency of the precedence relationship throughout the network. One way of doing so is to develop an *adjacency matrix*, which is a $n \times n$ matrix where n is the number of nodes in the network. The matrix has an entry $+1$ for element (ij) if an arrow leads from node i to j, and 0 otherwise. (Note that the nodes are numbered such that an arrow leads from a smaller numbered node to a larger one).

The precedence relationships of the project network can be represented by directed graphs or Boolean matrices. The implicit form of inconsistency or contradications in the graphical form is a closed loop of arrows head to tail,

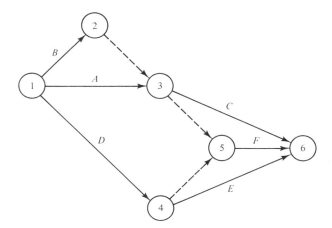

Figure 6-8. Correct representation of Example 6-2.

or, in general, a subset every member of which is a link. This implies that an activity may precede itself. The matrix of a set of items in a closed loop will contain an element in every row and every column; that is, there will be no zero rows or zero columns. Since the matrix of a subset of the original set of items will be a principal submatrix of the original adjacency matrix, we conclude that a set of precedence relations is consistent if and only if every principal submatrix has at least one zero row or zero column. This method for checking consistency was suggested by Marimont [16]. We now summarize the steps for checking the consistency of the network.

1. Delete every row and column of zeros and the same numbered column or row; that is, if row (column) i is deleted because it is all zeros, also delete column (row) i. This step will yield a new submatrix with new sources and terminals—namely, those nodes previously linked to the (now deleted) previous sources and terminals.
2. Repeat step 1 until either:
   (a) Every line in the matrix has been deleted, indicating a consistent precedence relationship, or
   (b) A submatrix with no zero lines remains, indicating an inconsistent set.

The following examples exhibit the construction and use of an adjacency matrix to check the consistency of a project network.

EXAMPLE 6-3

Construct a project network with the following precedence relationships and check for the consistency of the network.

$$A < F \qquad B < C, D, E$$
$$D < I \qquad F < I$$
$$E < J \qquad C < G, H$$
$$J < K \qquad H < K$$
$$G < J$$

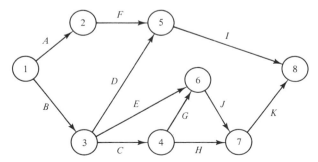

Figure 6-9. Network for Example 6-3.

SOLUTION

Figure 6-9 represents the project network as described by the relationships shown above. The adjacency matrix of this network is

	1*	2	3	4	5	6	7	8
1		1	1					
2				1				
3				1	1	1		
4						1	1	
5								1
6							1	
7								1
8*								

Delete columns and rows 1* and 8*.

	2*	3*	4	5	6	7
2			1			
3			1	1	1	
4					1	1
5*						
6						1
7*						

Delete columns and rows 2*, 3*, 5*, and 7*.

	4*	6
4		1
6*		

Now every line in the matrix has been deleted, indicating the consistency of the network.

6.2.7. CRITICAL PATH ALGORITHM

The *critical path* of a project network is the longest path through the network from the source to the terminal. The critical path defines the total duration of the project. In this section we present an algorithm for finding the critical path of a project and the start and finish times of all activities within the project. The following definitions will be used in the algorithm.

1. *The early start time (ES):* The early start time of a job in a project is the earliest possible time that the job can begin. The early start of job *A* is written as ES(*A*).
2. *The early finish time (EF):* The early finish time of a job is its early start time plus the duration of the job (time required to complete the

job). The early finish time of job A is labeled EF(A). Thus EF(A) = ES(A) + t_A, where t_A is the duration of job A.

In the following example, we present how ES and EF are estimated for all jobs in the network.

EXAMPLE 6-4

Estimate ES and EF for all jobs of the network given in Fig. 6-10.

Activity	Duration
A	5
B	4
C	8
E	6
F	4
G	5
H	2
I	3
J	4

SOLUTION

Before a job could be started, all its immediate predecessors must have been completed. For jobs with no predecessors, the start time of the job is set to zero. For example, the start time for jobs A, B, and C is zero. Therefore,

$$\text{ES}(A) = \text{ES}(B) = \text{ES}(C) = 0$$

and the corresponding early finish times are

$$\text{EF}(A) = \text{ES}(A) + t_A$$

$$\text{EF}(A) = 0 + 5 = 5$$

$$\text{EF}(B) = 0 + 4 = 4$$

$$\text{EF}(C) = 0 + 8 = 8$$

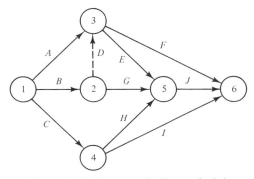

Figure 6-10. Network for Example 6-4.

Now, let us look at job E. It cannot start until immediate predecessors A and D are completed. But job D cannot start until B is completed; then the $ES(D)$ is the same as the $EF(B)$. Therefore,

$$EF(D) = ES(D) + t_D = 4 + 0 = 4$$

Now we can go back to job E. The $EF(A)$ and $EF(D)$ are 5 and 4, respectively. Therefore, $ES(E)$ equals the longest $EF(\text{job } i)$ of all the jobs preceding E.

$$ES(E) = 5 \quad \text{and} \quad EF(E) = 5 + 6 = 11$$

Also,

$$ES(F) = ES(E) = 5 \quad \text{and} \quad EF(F) = 5 + 4 = 9$$

Job B must be completed before G can start.

$$ES(G) = EF(B) = 4$$
$$EF(G) = 4 + 5 = 9$$

Similarly, jobs H and I cannot start before C is completed.

$$ES(H) = ES(I) = EF(C) = 8$$
$$EF(H) = 8 + 2 = 10$$
$$EF(I) = 8 + 3 = 11$$

Job J cannot start until jobs E, G, and H are completed.

$$ES(J) = \max \{EF(E), EF(G), EF(H)\}$$
$$ES(J) = \max \{11, 9, 10\}$$
$$ES(J) = 11 \quad \text{and} \quad EF(J) = 11 + 4 = 15$$

If the earliest completion date (max $\{EF(i)\} \; \forall \; i$) of the project is 15 days after its beginning, the longest path through its network must be 15 days in length. The path consisting of jobs A, E, and J is the longest path in the network. This sequence is called the *critical path of the network*, and jobs A, E, and J are called *critical activities*.

3. *Latest start time (LS):* It is obvious that activities that are not on the critical path can be delayed without delaying the completion time (CT) of the project. How much can an activity be delayed without affecting the completion time of the project? *Late start* is the latest time that an

activity can start without pushing the completion date of the entire project into the future.

$$LS(A) = LF(A) - t_A$$

where $LF(A)$ is the *late finish* of activity A. It is defined as the late start time of the activity plus its duration.

$$LF(A) = LS(A) + t_A$$

In order to calculate LS and LF, we start from the terminal node (sink) of the network and work backward. We set CT (completion time) to be LF of the project. At node 6, jobs F, J, and I should be completed to meet the project date of 15 days; in other words, day 15 is the project's late finish.

$$LF(F) = LF(J) = LF(I) = 15$$

Since it takes 4 days to complete job J, it must begin no later than $15 - 4 = 11$ days, which is its late start.

$$LS(J) = 11$$
$$LS(F) = 15 - 4 = 11$$
$$LS(I) = 15 - 3 = 12$$

Activities E, G, and H must be completed by day 11 [which is the $LS(J)$]. Consequently,

$$LF(E) = LF(G) = LF(H) = LS(J) = 11$$
$$LS(E) = 11 - 6 = 5$$
$$LS(G) = 11 - 5 = 6$$
$$LS(H) = 11 - 2 = 9$$

We now move to node 3. The LS of all jobs leaving this node must be compared and the minimum LS represents the LF of predecessor jobs:

$$LF(A) = LF(D) = \min \{LS(F), LS(E)\}$$
$$LF(A) = LF(D) = \min \{11, 5\} = 5$$
$$LS(A) = 5 - 5 = 0$$
$$LS(D) = 5 - 0 = 5$$

At node 2, we compare LS(G) and LS(D) to determine the LF(B).

$$LF(B) = \min \{LS(G), LS(D)\}$$
$$LF(B) = \min \{6, 5\} = 5$$
$$LS(B) = 5 - 4 = 1$$

Similarly, at node 4,

$$LF(C) = \min \{LS(H), LS(I)\}$$
$$LF(C) = \min \{9, 12\} = 9$$
$$LS(C) = 9 - 8 = 1$$

Finally, comparing LS(A), LS(B), and LS(C) and selecting the shortest time, we obtain

$$LS \text{ (at node 1)} = \min \{LS(A), LS(B), LS(C)\}$$
$$LS \text{ (at node 1)} = \min \{0, 1, 1\} = 0$$

ES, LS, EF, and LF of the activities in the network are shown in Fig. 6-11.

All activities that have ES = LS and EF = LF are critical activities and they constitute the critical path. Looking at Fig. 6-11, we observe that for some of the jobs (critical jobs) ES = LS, and EF = LF. For example, ES(A) = LS(A) = 0, and EF(J) = LF(J) = 15. On the other hand, some jobs, such as F, have ES(F) = 5 while LS(F) = 11. This means that job F may start any time between day 5 and day 11 without delaying the completion date of

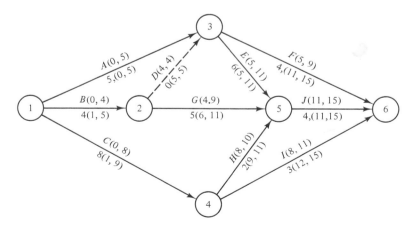

Figure 6-11. ES, EF, LS, and LF activities. Job name (ES, EF)/duration (LS, LF).

the project. We then say that job F has a *slack* (*float*) time. *Total slack* (TS) of a job or activity is the difference between its late start and early start times (or the difference between its late finish and early finish times).

$$TS(B) = LS(B) - ES(B) = 1 - 0 = 1$$

or

$$TS(B) = LF(B) - EF(B) = 5 - 4 = 1$$

It is obvious that the critical jobs or activities have total slack values of zero.

Introduced now, is a new definition, *free slack* (FS). From Fig. 6-11 one can notice that job J cannot start until job E is completed on day 11. Jobs G and H are also predecessors of job J, but they could be completed on days 9 and 10, respectively. Thus job G can start as late as day 6, while job H can start as late as day 9 without delaying the early start time of job J. We say that jobs G and H have free slack (FS) time, which we define as the amount of time a job can be delayed without affecting the early start time of other jobs. FS of a job is the difference between its early finish time and the earliest of the early start times of all its immediate successors. Thus

$$FS(E) = ES(J) - EF(E) = 11 - 11 = 0$$

$$FS(G) = ES(J) - EF(G) = 11 - 9 = 2$$

$$FS(H) = ES(J) - EF(H) = 11 - 10 = 1$$

Note that the free slack can never exceed total slack.

A convenient procedure to find the critical path is the *tabulation method*. This method is illustrated by Table 6-1. Activities A, E, and J have total slack times of zero. Thus A, E, and J are critical activities.

Another approach for finding the critical path is the *linear programming* approach. In using linear programming, we view the project network as a flow network in which a unit flow enters at origin 1 and exits at the terminal node

Table 6-1. TABULATION METHOD FOR CRITICAL PLAN

Activity	Duration, t	ES	LS	EF (ES + t)	LF (LS + t)	TS (LF − EF)
A	5	0	0	5	5	0
B	4	0	1	4	5	1
C	8	0	1	8	9	1
D	0	4	5	4	5	1
E	6	5	5	11	11	0
F	4	5	11	9	15	6
G	5	4	6	9	11	2
H	2	8	9	10	11	1
I	3	8	12	11	15	4
J	4	11	11	15	15	0

n. The duration of each activity, t_{ij}, is the time of transportation of the commodity from node i to node j. Under such modeling, determining the critical path (longest path in the network) is equivalent to determining the path of the maximum flow from node 1 to node n.

6.2.8. LINEAR PROGRAMMING FORMULATION

Let $x_{ij} \geq 0$ be the quantity of the commodity flowing in the arc (ij) in the direction $i \rightarrow j$. Then the primal linear programming is given by

$$\text{Maximize} \sum_{\forall ij} t_{ij} x_{ij} \tag{6.1}$$

subject to

$$x_{ij} \geq 0 \tag{6.2}$$

$$\sum_{j} x_{1j} = 1 \qquad \forall \ 1j \tag{6.3}$$

$$-\sum_{i} x_{ik} + \sum_{j} x_{kj} = 0 \qquad \forall \ k \neq 1, n \tag{6.4}$$

$$-\sum_{i} x_{in} = -1 \qquad \forall \ in \tag{6.5}$$

where node 1 is the source node and node n the terminal node.

Constraint (6.3) restricts the 1 unit of flow which enters the source (node 1) to flow only over one of the arcs leaving this source node. Constraint (6.4) represents the conservation of flow at any node (in flow = out flow). While constraint (6.5) states that only 1 unit of flow leaves the terminal node. The solution of the linear programming problem yields the values x_{ij} for each arc. The critical path consists of arcs whose $x_{ij} = 1$.

We now apply the linear programming technique to the network given in Example 6-4.

EXAMPLE 6-5

Use the linear programming approach to find the critical path for the network given in Example 6-4.

SOLUTION

The objective function of the linear programming problem is given as

$$\text{Maximize } 4x_{12} + 5x_{13} + 8x_{14} + 0x_{23} + 5x_{25} + 6x_{35}$$
$$+ 4x_{36} + 2x_{45} + 3x_{46} + 4x_{56} \tag{6.6}$$

subject to

$$x_{12} + x_{13} + x_{14} \qquad \qquad = 1 \tag{6.7}$$

$$-x_{12} \qquad \qquad + x_{23} + x_{25} \qquad \qquad = 0 \tag{6.8}$$

$$-x_{13} \qquad -x_{23} \qquad +x_{35}+x_{36} \qquad\qquad\qquad =0 \qquad (6.9)$$

$$-x_{14} \qquad\qquad\qquad +x_{45}+x_{46} \qquad =0 \qquad (6.10)$$

$$-x_{25}-x_{35} \qquad -x_{45} \qquad +x_{56}=0 \qquad (6.11)$$

$$-x_{36} \qquad -x_{46}-x_{56}=-1 \qquad (6.12)$$

and $x_{ij} \geq 0$.

The dual of this linear programming problem can be presented as follows. Let $y_i =$ dual variable corresponding to the ith equation in the primal problem. Therefore, the dual problem becomes

$$\text{Minimize } y_1 - y_6 \qquad\qquad (6.13)$$

subject to

$$y_1 - y_2 \qquad\qquad \geq 4 \qquad\qquad (6.14)$$

$$y_1 \qquad - y_3 \qquad\qquad \geq 5 \qquad\qquad (6.15)$$

$$y_1 \qquad\quad - y_4 \qquad\qquad \geq 8 \qquad\qquad (6.16)$$

$$y_2 - y_3 \qquad\qquad \geq 0 \qquad\qquad (6.17)$$

$$y_2 \qquad\quad - y_5 \qquad \geq 5 \qquad\qquad (6.18)$$

$$y_3 \qquad - y_5 \qquad \geq 6 \qquad\qquad (6.19)$$

$$y_3 \qquad\qquad - y_6 \geq 4 \qquad\qquad (6.20)$$

$$y_4 - y_5 \qquad \geq 2 \qquad\qquad (6.21)$$

$$y_4 \qquad - y_6 \geq 3 \qquad\qquad (6.22)$$

$$y_5 - y_6 \geq 4 \qquad\qquad (6.23)$$

Since each inequality in the dual problem contains two and only two variables, the problem can be solved by inspection. One solution of this dual problem is

$$y_1 = 15, \quad y_2 = 10, \quad y_3 = 10, \quad y_4 = 6, \quad y_5 = 4, \quad y_6 = 0$$

The optimal dual solution is $y_1 - y_6 = 15$ (which represents the length of the critical path). Since the inequalities (6.15), (6.17), (6.19), and (6.23) are satisfied as equalities, the variables corresponding to these inequalities in the primal problem are set to be ≥ 0 (primal–dual relationship); that is, $x_{13}, x_{23}, x_{35}, and \; x_{56} \geq 0$. Substituting other values in Eqs. (6.6) through (6.11) equal to zero, we obtain

$$x_{13} \qquad\qquad\qquad =1$$

$$x_{23} \qquad\qquad =0$$

$$-x_{13}-x_{23}+x_{35} \qquad =0$$

$$-x_{35}+x_{56}=0$$

$$-x_{56}=-1$$

Solution of the equations above yield

$$x_{13} = x_{35} = x_{56} = 1 \qquad x_{23} = 0$$

Thus the critical path is $1 \to 3 \to 5 \to 6$, with a total duration of 15 days.

6.2.9. Cost Models

In developing the critical path in the previous sections, we did not take the cost of project activities into account. In this section we present various cost models that can be utilized in project planning and control. When considering cost, the project manager will always seek the answer to one or more of the following questions:

1. What is the optimal cost duration curve for the project?
2. Given a due date for the project, what is the set of decisions that will minimize the cost of meeting the due date?
3. Given a fixed budget for the project, what is the set of decisions that will minimize the completion time of the project under this budget constraint?

Historical experience shows that the cost of a project can be accurately estimated, can be closely monitored, and can be controlled. One method of monitoring and controlling the project cost is by accumulating actual cost data with time. These data can be collected monthly, weekly, or daily: the frequency of cost data collection will depend on the type of project and the cost to prepare and present such data. Each project will have a unique cost–time function based on the type of project and type of contract. It should be recognized that under a specific type of contract wherein the project must be completed by a certain due date, the contract will often specify penalties for delays beyond this date. It was reported that Lockheed's contract for the C-5A aircraft contained a $12,000 per day late penalty clause and that General Dynamics was rewarded $800,000 for flying the F-16 aircraft 10 days ahead of schedule.

The critical path method should be used where it is possible to predict time durations for activities and costs within a very close range. It assumes that the total project costs are the sum of direct and indirect costs. As presented earlier, the initial step in CPM is to develop the network diagram. Next, the duration and cost of each activity are estimated. Most CPM procedures utilize two time–cost estimates. These two estimates are called the *normal* and *crash* time–cost estimates. We define these estimates as follows:

1. *Normal time and cost:* The duration to complete the project requiring the least amount of money

2. *Crash time and cost:* The minimum possible time to complete the activity and the associated increased cost

Suppose that we are asked to crash some activities in order to decrease the duration of the project. One might ask: Which activities should be crashed? and by how much? The answer to the first part is obvious: one should crash the activity that has minimum crashing cost per unit time and is on the critical path. The answer to the second part of the question is a function of the normal and crash time–cost of all activities.

We present a methodology for crashing activities in order to minimize the increase in the total cost of the project.

—*Step 1:* Calculate the cost change per unit time (CC/UT).

$$CC/UT = \frac{\text{crash cost} - \text{normal cost}}{\text{normal time} - \text{crash time}} = \$/\text{unit time}$$

—*Step 2:* Determine the project normal schedule as presented earlier.
—*Step 3:* Determine the critical path (CP) for the normal schedule.
—*Step 4:* Determine the direct cost of the normal schedule as the sum of the normal costs for all activities.
—*Step 5:* Reduce (crash) the schedule by one time unit. Only those activities on the critical path are considered for crashing (crashing activities not on the critical path will not reduce the length of the critical path). Only one activity on each critical path is crashed. (The activity on each critical path that is to be crashed is the one that will achieve a minimum CC/UT for all paths to be crashed.)
—*Step 6:* Repeat step 5 until there are no more jobs to be crashed.

EXAMPLE 6-6

Consider the network given in Example 6-4. The following are normal and crash times and the costs of all activities. Find the crashing schedule that minimizes the total project cost.

	Time (days)		Cost	
Activity	Normal	Crash	Normal	Crash
A	5	3	$100	$120
B	4	3	110	130
C	8	5	95	110
E	6	4	80	90
F	4	3	150	180
G	5	4	90	130
H	2	2	70	70
I	3	3	100	100
J	4	2	200	220

SOLUTION

$$CC/UT = \frac{\text{crash cost} - \text{normal cost}}{\text{normal time} - \text{crash time}}$$

Activity	CC/UT ($/unit time)
A	10
B	20
C	5
E	5
F	30
G	40
H	No
I	No
J	10

Direct cost for the normal schedule is $995. The critical path is shown in Fig. 6-12. This critical path is crashed by 1 day to be 14 days. The activity that has the minimum CC/UT among *A*, *E, and J* is the one that should be crashed.

CC/UT for *A* = $10

CC/UT for *E* = $ 5

CC/UT for *J* = $10

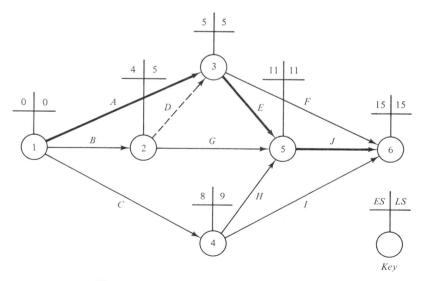

Figure 6-12. The critical path of Example 6-6.

Table 6-2. MINIMUM-COST CRASHING SCHEDULE

Iteration	Critical Path	Job Crashed and (New Time)	Added Direct Cost	Total Direct Cost	Project Time	Activities at Minimum Time
0	AEJ	—	—	$ 995	15	H, I
1	AEJ CHJ	E(5)	5	1000	14	H, I
2	AEJ CHJ BGJ	E(4) C(7)	5 5	1010	13	H, I, E
3	AEJ CHJ BGJ	J(3)	10	1020	12	H, I, E
4	AEJ CHJ BGJ	J(2)	10	1030	11	H, I, E, J
5	AEJ CHJ BGJ	A(4) C(6) B(3)	10 5 20	1065	10	H, I, E, J, B
6	AEJ CHJ BGJ	A(3) C(5) G(4)	10 5 40	1120	9	H, I, E, J, B A, C, G

We crash activity E by 1 day. This will increase the total cost by $5; project cost then becomes $1000 and the project will be completed in 14 days. We continue crashing activities on the critical paths until we are unable to crash any more activities. Table 6-2 illustrates this procedure.

The crashing schedule that minimizes the project cost is 9 days at a cost of $1120.

6.2.10. PROGRAM EVALUATION AND REVIEW TECHNIQUE (PERT)

Both CPM and PERT use the project network. CPM assumes that the durations of activities are deterministic; PERT allows for uncertainties in activity times. In PERT, we replace the single value of activity duration estimate in CPM with three estimates of time, and the mean flow time for each activity is calculated using these three estimates of time. In this section we present a methodology for computing the expected flow time of a project.

After constructing the network, three time estimates are made for each activity: the *optimistic*, *pessimistic*, and *most likely*. These estimates are obtained by asking the following questions:

1. What is the least amount of time to complete an activity under the best possible circumstances? *Optimistic time estimate.*

2. What is the greatest amount of time that would be required to complete the activity under the worst possible circumstances? *Pessimistic time.*

3. What is the most probable amount of time in which the activity could be completed under normal circumstances? *Most likely time.*

These three time estimates for each activity are then reduced to a single value to be used in computing the project duration. We define t_a as the optimistic time estimate, t_b as the pessimistic time, and t_m is the most likely estimate. It is assumed that t_a and t_b are equally likely to occur and the most probable activity time t_m is four times more likely to occur than either of the other two. This characterizes a beta probability distribution. If we apply these weights to the three time estimates, we obtain the expected time t_e of an activity as

$$t_e = \frac{t_a + 4t_m + t_b}{6} \qquad (6.24)$$

After estimating t_e for each activity in the network, one can then apply the CPM algorithm in order to obtain the critical path(s) in the network.

Since there are uncertainties in the time estimates of the activities, we would be interested in knowing how reliable these estimates are. If the time required for an activity is highly variable—if the range of estimates is very large—then we have less confidence in the expected time value than if we have a narrower range of estimates. One measure of this is the standard deviation of the probability distribution of activity time. In PERT, it is assumed that t follows the beta distribution. Therefore, the standard deviation of an activity time is estimated by

$$\sigma_t = \frac{t_b - t_a}{6}$$

The expected project length, CT (completion time), is calculated by adding the t_e's along the critical path. Since t_e's are assumed to be independently distributed, the central limit theorem implies that as the number of activities along the critical path increases, CT approaches the normal distribution. If we are interested in determining the probability that the project will be completed by the due date D, we calculate Z, the number of standard deviations by which D exceeds CT. Note that D might be less than CT, in which case Z will have a negative value. The formula for calculating Z is as follows:

$$Z = \frac{D - CT}{\sigma_{\text{critical}}}$$

where D = due date

CT = completion time of the project

$\sigma_{\text{critical}} = \sqrt{\sum \sigma^2}$ (of all activities on the critical path)

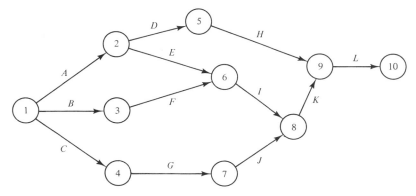

Figure 6-13. Project network for Example 6-7.

Using the value of Z, one can obtain the probability from the table of the standard normal probability density function.

EXAMPLE 6-7

The following are the time estimates of all activities in the network given in Fig. 6-13. Find the critical path CT (*completion time* of the project) and the probability that the project will be completed in (CT + 5) days.

	Time Estimates		
Activity	t_a	t_m	t_b
A	4	5	6
B	5	6	8
C	3	4	7
D	4	5	6
E	3	3	3
F	3	4	8
G	5	8	12
H	1	1	1
I	1	2	3
J	2	3	6
K	4	7	9
L	2	5	7

SOLUTION

We calculate t_e for each activity as

$$t_{e_i} = \left(\frac{t_a + 4t_m + t_b}{6}\right)_i \qquad \text{for activity } i$$

$$\sigma_i^2 = \left(\frac{t_b - t_a}{6}\right)_i^2$$

Table 6-3. t_e AND σ^2 FOR
ACTIVITIES

Activity	t_{e_i}	σ_i^2
A	5.00	0.111
B	6.17	0.250
C	4.33	0.444
D	5.00	0.111
E	3.00	0.000
F	4.50	0.694
G	8.17	1.361
H	1.00	0.000
I	2.00	0.111
J	3.33	0.444
K	6.83	0.694
L	4.83	0.694

These values are listed in Table 6-3. The critical path is $C-G-J-K-L$ with a duration of 27.48 days. The probability that the project will be completed in $27.48 + 5 = 32.48$ days is calculated as follows:

$$Z = \frac{32.48 - 27.48}{\sqrt{0.44 + 1.36 + 0.44 + 0.69 + 0.69}} = \frac{5}{\sqrt{3.62}} = \frac{5}{1.902}$$

$$= 2.6288$$

From the tables of the standard normal distribution, the probability corresponding to Z of 2.6288 is 99.5%, which indicates that the probability of completing the project in 32.48 days is $1 - 0.995 = 0.005$.

6.2.11. LIMITATIONS OF PERT AND/OR CPM

Some of the limitations of PERT and/or CPM are as follows:

1. PERT is primarily restricted to unique large-scale projects. It is not used in the scheduling of production-type operations.
2. In PERT/CPM the actual performance of an activity may alter the entire network timing. Therefore, the original network is likely to change shortly after the beginning of the project.
3. In PERT and CPM, the times are estimated and therefore subject to the weaknesses of all estimating procedures.
4. In PERT, the assumption of a beta distribution for all activities and the assumption that enough observations exist along the critical path to use the estimation properties of the normal distribution is suspect.
5. In PERT, the assumption that activity completion times are uncorrelated is unlikely in many real situations.
6. All activities leading to an event must be completed before the event can be realized; in other words, partial realization is not allowed.

6.3. Project Planning and Scheduling: Limited Resources

Imposing capacity or resources restriction of some kind on project activities will lead to an increased completion time of the project. Consider the network given in Fig. 6-13. Let us assume that limited resources are available for the activities, more specifically, each activity requires the utilization of jib cranes. For example, activities *A*, *B*, *C*, and *D* require three cranes for each activity to be completed while activities *E*, *F*, and *G* require two cranes for each activity. Four cranes are needed for each of the following activities: *H*, *I*, *J*, *K*, and *L*. Assume we only have four cranes available. We would like to schedule these activites and find the optimal crane assignment in order to minimize the completion time of the project. Obviously, the completion time under no constraints is less than when constraints are imposed. Optimal solutions to this problem are very difficult to obtain, especially for large-scale networks.

One way of approaching this problem is to use *heuristic solution techniques*. A heuristic is a set of decision rules that might lead to an optimum solution but does not guarantee optimality. In other words, the heuristic solution may result in an optimum solution for some problems and may not for other problems. In the following sections, heuristic algorithms are presented that have been shown to achieve optimal or near-optimal solutions for project scheduling under limited resources.

6.3.1. ROT ALGORITHM

Elsayed [6] developed a heuristic algorithm for assigning resources to activities such that project duration is minimized. The criterion for allocating a single type of resource to the activites in the network is based on ROT (*resources over time*), which is calculated as the maximum ROT value that an activity controls through the network on any one path. This is similar to calculating the critical path time through the network assuming that the starting node for each activity being considered is the network starting node. For example, in the network given in Fig. 6-14, ROT value for activity *A* is 1.291, which is obtained as

$$\text{ROT}(A) = \max \left\{ \tfrac{2}{4} + \tfrac{4}{6} + \tfrac{1}{8}, \tfrac{2}{4} + \tfrac{2}{10} + \tfrac{1}{2} \right\} = \tfrac{31}{24} = 1.291$$

While ROT for activity *D* is 19/24.

After determining ROT values for each activity, we rank the activities in a decreasing ROT sequence and ties can be broken arbitrarily. TEARL of an activity is calculated as the earliest time to schedule this activity under the precedence and time constraints. In other words, TEARL of an activity is the latest TFIN (finish time) for *all immediate predecessor activities*. We define TSTART as the actual start time of the activity if the resources, time, and

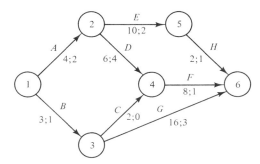

Figure 6-14. Project network with resource constraints. Time; resources.

precedence restrictions are taken into account. If the resource constraint is not considered, then TEARL = TSTART. In the example to follow, we will demonstrate how TSTART is computed. We also define TFIN and TNOW. TFIN is the completion time of each activity; that is, it is the sum of TSTART and the duration of the activity. TNOW is the time at which assignments are now being considered. At the beginning, we assign a value of zero to TNOW, but subsequently, it will be set as the minimum TFIN. The following example illustrates the use of ROT criterion in allocating resources to activities.

EXAMPLE 6-8

Find the critical path of the network shown in Fig. 6-14 without consideration for resource needs. Then utilize ROT algorithm to determine the minimum length of the project allowing 4 units of resource.

SOLUTION

Using the tabulation procedure, the critical path is found as shown in Table 6-4. From the table we see that the critical path is 1–3–6 and its duration is 19 days.

ROT algorithm is now utilized to assign the four available resources to the activities of the network. The steps of ROT algorithm are summarized as follows:

1. Determine for each activity the maximum ROT value it controls through the network on any one path. ROT values for the activities of this example are:

Activity	ROT
1–2	1.291
1–3	0.520
2–4	0.791
2–5	0.700
3–4	0.125
3–6	0.187
4–6	0.125
5–6	0.500

Table 6-4. CRITICAL PATH CALCULATIONS FOR EXAMPLE 6-8

Activity	Duration	ES_i	LS_i	EF_i	LF_i	TS_i
1–2	4	0	1	4	5	1
1–3	3	0	0	3	3	0
2–4	6	4	5	10	11	1
2–5	10	4	7	14	17	3
3–4	2	3	9	5	11	6
3–6	16	3	3	19	19	0
4–6	8	10	11	18	19	1
5–6	2	14	17	16	19	3

2. Rank activities in decreasing ROT sequence. Ties are broken by ranking the activity of longest duration first as shown in Table 6-5.
3. We start with TNOW = 0 and resources available = 4. Activities allowed that could start at TNOW = 0 are 1–2 and 1–3 and they are entered in Table 6-5 according to their ROT values.
4. Assign two of the resources to activity 1–2 and one of the resources to activity 1–3 (both require a total of three resources). Enter actual start times for these activities as zero in Table 6-5. TFIN for these activities is calculated by adding TSTART and the duration of the activity. This results in a TFIN of 4 and 3 for activities 1–2 and 1–3, respectively. There is still one unit of resource available, but it could not be assigned to either of the next activities to be scheduled (3–6 and 3–4) since both of them require the completion of activity 1–3.
5. We now move to the next iteration with a new value of TNOW, which is the minimum of the available TFIN. The TFIN available at this iteration are 3 and 4 and TNOW is set to be TNOW = min {3, 4}, or TNOW = 3.
6. There are two resources available at TNOW = 3, one of which was not used at

Table 6-5. ROT ALGORITHM FOR EXAMPLE 6-8

	Activity							
	1–2	2–4	2–5	1–3	5–6	3–6	4–6	3–4
ROT	1.291	0.791	0.70	0.520	0.50	0.187	0.125	0.125
Duration	4	6	10	3	2	16	8	2
Resources required	2	4	2	1	1	3	1	0
TEARL	0	4	4	0	14	3	10	3
TSTART	0	4	10	0	20	20	10	3
TFIN = TSTART + duration	4	10	20	3	22	(36)	18	5
TNOW	0	3	4	5	10	18	20	
Resources available	4̶ 2̶ 1	2	4̶ 0	0	4̶ 2̶ 1	2	4̶ 3̶ 0	
Activity allowed	1̶2, 1̶3	3–6, 3̶4	2̶4, 2–5, 3–6	2–5, 3–6	2̶5, 3–6, 4̶6	3–6	5̶6, 3̶6	
Iteration number	1	2	3	4	5	6	7	

TNOW $= 0$ and one is released after the completion of activity 1–3. Activities allowed—that is, those ready for scheduling—are 3–6 and 3–4 as their predecessor is completed (activity 1–3) and they have the minimum value of TEARL. Since activities 3–6 and 3–4 require 3 and 0 units of resources, respectively, and resources on hand at TNOW $= 3$ are only 2, we cannot assign these resources to activity 3–6 and activity 3–4 can start immediately, or TSTART for activity 3–4 is 3 and its TFIN $= 5$.

7. The next TNOW is the minimum of the available TFIN (4 and 5). TNOW is set to be 4 with four available resources (two are released from activity 1–2, and two are on hand). Again we determine the list of activities allowed based on ROT ranking, TEARL, and precedence relationship. Activities allowed at this iteration are 2–4, 2–5, and 3–6. We assign four resources to activity 2–4, since it has the largest ROT value and its TSTART $= 4$ while its TFIN $= 10$. No other resources are available and this completes iteration 3.

8. Repeat step 7, until all activities are scheduled.

The minimum completion time of this project under the resource constraints is the maximum TFIN, or 36 units of time.

Allocating resources to the activities based on ROT criterion does not guarantee optimality. In fact, we shall use other criteria later to obtain less time for completing this project than using ROT. In the following sections, we introduce other criteria for allocating resources to activities.

6.3.2. ACTIM Criterion[1]

The steps of allocating resources under the ACTIM (activity time) criterion are based on Brook's algorithm (BAG) and similar to those used under ROT except that the ranking of the activities in the table is dependent on the ACTIM value of the activity. The ACTIM value of an activity is calculated as the maximum time that the activity controls through the network on any one path. Table 6-6 illustrates the use of the ACTIM criterion in determining the project completion time of the network given in Example 6-8. Using the ACTIM criterion, the minimum completion time of this project is 38 days.

Activity	ACTIM
1–2	18
1–3	19
2–4	14
2–5	12
3–4	10
3–6	16
4–6	8
5–6	2

---

[1] Materials for the ACTIM, ACTRES, and TIMRES criteria are based on Ref. 4.

Table 6-6. PROJECT SCHEDULING USING ACTIM

	Activity							
	1–3	*1–2*	*3–6*	*2–4*	*2–5*	*3–4*	*4–6*	*5–6*
ACTIM	19	18	16	14	12	10	8	2
Duration	3	4	16	6	10	2	8	2
Resources required	1	2	3	4	2	0	1	1
TEARL	0	0	3	4	4	3	10	14
TSTART	0	0	4	20	26	3	26	36
TFIN	3	4	20	26	36	5	34	⑧
TNOW	0	3	3	5	20	26	34	36
Resources available	4̶ 3̶ 1	2	4̶ 1	1	4̶ 0	4̶ 2̶ 1	2	4̶ 3
Activity allowed	1↓3, 1↓2	3–6, 3↓4	3↓6, 2–4, 2–5	2–4, 2–5	2↓4, 2–5	2↓5, 4↓6	5–6	5↓6
Iteration number	1	2	3	4	5	6	7	8

6.3.3. ACTRES AND TIMRES CRITERIA

The steps of allocating resources to activities based on the ACTRES (activity resources) criterion are similar to those of ACTIM with the exception that the ranking of the activities is based on their ACTRES values (Bedworth [2]). We calculate ACTRES by multiplying each activity's time by its resources and then finding the maximum ACTRES that an activity controls through the network on any one path. We now use the ACTRES criterion to estimate the project duration of the network given in Example 6-8.

As shown in Table 6-7, the project completion time under the ACTRES

Table 6-7. ACTRES CRITERION FOR EXAMPLE 6-8

	Activity							
	1–3	*3–6*	*1–2*	*2–4*	*2–5*	*3–4*	*4–6*	*5–6*
ACTRES	51	48	40	32	22	8	8	2
Duration	3	16	4	6	10	2	8	2
Resources required	1	3	2	4	2	0	1	1
TEARL	0	3	0	4	4	3	10	14
TSTART	0	4	0	20	26	3	26	36
TFIN	3	20	4	26	36	5	34	㊳
TNOW	0	3	4	20	26	34	36	
Resources available	4̶ 2̶ 1	2	4̶ 1	4̶ 0	4̶ 2̶ 1	2	4̶ 3	
Activity allowed	1↓3, 1↓2	3–6, 3↓4	3↓6, 2–4, 2–5	2↓4, 2–5	2↓5, 4↓6		5↓6	
Iteration number	1	2	3	4	5	5	7	

Table 6-8. TIMRES VALUES FOR THE ACTIVITIES
OF EXAMPLE 6-8

Activity	ACTIM	ACTRES	TIMRES
1–2	18	40	58
1–3	19	51	70
2–4	14	32	46
2–5	12	22	34
3–4	10	8	18
3–6	16	48	64
4–6	8	8	16
5–6	2	2	4

criterion is the same as that obtained by ACTIM of 38 days. If we apply ACTIM and ACTRES to another project network, we may obtain different results and ACTIM might give a shorter project completion time than ACTRES, or vice versa.

Some combinations of criteria presented earlier are suggested. Mason [17] suggested a combination of ACTIM and ACTRES, which is designated as the TIMRES—time and resources combination. As shown in Table 6-8, the value of TIMRES of an activity is calculated as the sum of ACTIM and ACTRES values of this activity. We now apply TIMRES for the same problem given in Example 6-8. Again, as shown in Table 6-9, the completion time for the project using the TIMRES criterion is the same as the times for ACTIM and ACTRES.

6.3.4. MODIFICATION OF THE TIMRES CRITERION—GENRES

Whitehouse and Brown [30] modified the TIMRES criterion to obtain the least possible project completion time. The premise of their approach, called *GENRES*, is that further project schedules can be generated using com-

Table 6-9. PROJECT SCHEDULING USING TIMRES

				Activity				
	1–3	3–6	1–2	2–4	2–5	3–4	4–6	5–6
TIMRES	70	64	58	46	34	18	16	4
Duration	3	16	4	6	10	2	8	2
Resources required	1	3	2	4	2	0	1	1
TEARL	0	3	0	4	4	3	10	14
TSTART	0	4	0	20	26	3	26	36
TFIN	3	20	4	26	36	5	34	⃝38
TNOW	0	3	4	5	20	26	34	36
Resources available	4̶ 3̶ 1	2	4̶ 1	1	4̶ 0	4̶ 2̶ 1	2	4̶ 3
Activity allowed	1̶3, 1̶2	3–6, 3̶4	3̶6, 2–4, 2–5	2–4, 2–5	2̶4, 2–5	2̶5, 4̶6	5–6	5̶6
Iteration number	1	2	3	4	5	6	7	8

binations of ACTIM and ACTRES which are not equally weighted. Various weightings would be tried and the best project schedule is selected as given in ACTIM. The GENRES criterion is calculated as

$$\text{GENRES} = (W)(\text{ACTRES}) + (1 - W)(\text{ACTIM}) \qquad (6.25)$$

where W is an assigned weight $(0 \le W \le 1)$. The values of GENRES and TIMRES are equal at $W = 0.5$. A flowchart for using the GENRES criterion for project scheduling as developed by Whitehouse and Brown [30] is shown in Fig. 6-15.

The steps of the GENRES algorithm are summarized as:

1. Assign a weight W $(0 \le W \le 1)$ to calculate the GENRES values of each activity according to Eq. (6.25). We usually start with $W = 0$.

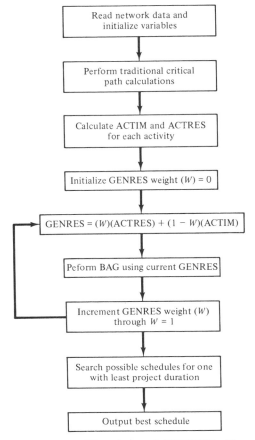

Figure 6-15. Flowchart of GENRES. (Reprinted by permission of Pergamon Press.)

2. Rank activities in decreasing order according to their GENRES values.
3. The project duration is calculated for this weight using the same procedures of ROT, ACTIM, and ACTRES.
4. Repeat steps 1 through 3 with an incremented value of W.
5. The best project schedule is then chosen as the one with the least project duration.

6.3.5. MODIFICATION OF ROT (ROT-ACTIM, ROT-ACTRES)

In order to minimize the total project duration, Elsayed [6] proposed the following criteria, which are modifications of ROT.

1. ROT-ACTIM: The weighted sum of ROT and ACTIM. The value of ROT-ACTIM of any activity is given by

$$\text{ROT-ACTIM} = (W)(\text{ROT}) + (1 - W)(\text{ACTIM})$$

2. ROT-ACTRES: The weighted sum of ROT and ACTRES.

$$\text{ROT-ACTRES} = (W)(\text{ROT}) + (1 - W)(\text{ACTRES})$$

The procedures of assigning resources to activities is similar to those presented earlier. The results of using ROT-ACTIM and ROT-ACTRES criteria will be discussed later.

6.3.6. TMROS

Nasr [21] improved the criteria mentioned previously by proposing TMROS. TMROS utilizes two criteria: ACROS and ACTIM. The ACROS criterion of an activity is calculated as the maximum resource value it controls through the network on any path. The method of calculating ACROS for each activity is similar to that used in calculating ACTIM, but resource values are used instead of time values.

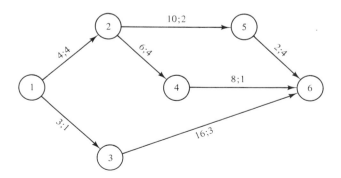

Figure 6-16. Network for Example 6-9. Time; resource.

The TMROS criterion is generated by combining different weights of ACROS and ACTIM using Eq. (6.26). The project completion time is then calculated for each weighted combination. Finally, the best project schedule is chosen as the one with the shortest project duration.

$$\text{TMROS} = (W)\text{ACROS} + (1 - W)\text{ACTIM} \qquad (6.26)$$

As Example 6-9 illustrates, the values of ACROS and ACTIM are scaled on a scale of 0 to 100.

EXAMPLE 6-9

Use TMROS to find the minimum project duration of the network given in Fig. 6-16 (the maximum number of available resources = 5).

SOLUTION

We first calculate ACTIM and ACROS values for each activity and scale it on a 0–100 scale, as shown in Table 6-10. We then illustrate the use of TMROS at $W = 0$ and $W = 0.5$, and the project schedules are shown in Tables 6-11 and 6-12, respectively. The project schedules for $W = 0$ and $W = 0.5$ are 34 and 28, respectively. The project schedules for different weights are summarized in Table 6-13.

6.3.7. TG2

The criterion TG2 is generated by combining different weights of TMROS and GENRES. The project completion time is then calculated for each weighted combination. Finally, the best project schedule is chosen as the one with the shortest project duration. The steps for using TG2 in project scheduling are:

1. Perform TMROS calculations until the best schedule is obtained.
2. Repeat step 1 using GENRES criterion and find the best schedule.
3. The criterion TG2 for each activity is estimated based on weighted

Table 6-10. ACTIM AND ACROS VALUES FOR EXAMPLE 6-9

Activity	ACTIM Value	ACTIM 0–100 Scale	ACROS Value	ACROS 0–100 Scale
1–2	18	94.74	10	100
1–3	19	100.00	4	40
2–4	14	73.68	5	50
2–5	12	63.16	6	60
3–6	16	84.21	3	30
4–6	8	42.11	1	10
5–6	2	10.53	4	40

Table 6-11. SCHEDULE OF THE NETWORK OF EXAMPLE 6-9 BY TMROS AT $W = 0$ AND TOTAL RESOURCES OF 5

	Activity						
	1–3	*1–2*	*3–6*	*2–4*	*2–5*	*4–6*	*5–6*
TMROS	100	94.74	84.21	73.68	63.16	42.11	10.53
Duration	3	4	16	6	10	8	2
Resources required	1	4	3	4	2	1	4
TEARL	0	0	3	4	4	10	14
TSTART	0	0	4	20	4	26	26
TFIN	3	4	20	26	14	(34)	28
TNOW	0	3	4	14	20	26	
Resources available	5 4 0	1	5 2 0	2	5 1	5 4 0	
Activity allowed	1+3, 1+2	3–6	3+6, 2–4, 2+5	2–4, 5–6	2+4, 5–6	4+6, 5+6	
Iteration number	1	2	3	4	5	6	

values of its **TMROS** and **GENRES** values in steps 1 and 2, respectively. In other words, the criterion TG2 of activity *i* is estimated as

$$(TG2)_i = W(TMROS)_i + (1 - W)(GENRES)_i$$

where $(TMROS)_i$ = TMROS value of activity *i* at the best schedule
$(GENRES)_i$ = GENRES value of activity *i* at the best schedule
W = assigned weight $(0 \leq W \leq 1)$

Table 6-12. SCHEDULE OF THE NETWORK OF EXAMPLE 6-9 BY TMROS AT $w = 0.5$ AND TOTAL RESOURCES OF 5

	Activity						
	1–2	*1–3*	*2–4*	*2–5*	*3–6*	*4–6*	*5–6*
TMROS	97.37	70.00	61.84	61.58	57.11	26.05	25.26
Duration	4	3	6	10	16	8	2
Resources required	4	1	4	2	3	1	4
TEARL	0	0	4	4	3	10	14
TSTART	0	0	4	10	10	20	26
TFIN	4	3	10	20	26	(28)	(28)
TNOW	0	3	4	10	20	26	
Resources available	5 1 0	1	5 1	5 3 0	2 1	4	
Activity allowed	1+2, 1+3	3–6	2+4, 2–5, 3–6	2+5, 3+6, 4–6	4+6, 5–6	5+6	
Iteration number	1	2	3	4	5	6	

Table 6-13. SUMMARY OF PROJECT DURATIONS FOR EXAMPLE 6-9 AT DIFFERENT WEIGHTS

W	0.00	0.10	0.20	0.30	0.40	0.50	0.60	0.70	0.80	0.90	1.00
Project *Duration*	34	34	34	34	28	28	34	34	34	34	34

4. Assign a weight W to calculate TG2 for each activity. Rank activities in a decreasing order according to their TG2 values.
5. The project duration is calculated for this weight.
6. Repeat steps 4 and 5 for different values of W. The best schedule is the one with the shortest project duration.

Experimentations with TG2 for different networks show that the use of TG2 improves the project schedule in large networks more often than in small networks. TMROS and GENRES exhaust all possible schedules. This implies that TG2 will give the best schedule in small networks at $W = 0$ or $W = 1$.

6.3.8. PERFORMANCE OF THE ALGORITHMS

The aforementioned algorithms, ROT, ACTIM, ACTRES, GENRES, ROT-ACTIM, ROT-ACTRES, TMROS, and TG2, are used in different networks and at different resource levels. A summary of the performance of these criteria is given in Table 6-14.

6.3.9. MULTIPLE-PROJECT, MULTIPLE-RESOURCE ALGORITHM

Scheduling of multiple projects with multiple resources is a complex problem. It usually takes enormous amount of computations to achieve a reasonable allocation of resources to activities. Most of the algorithms in the literature which deal with multiple-resource constraints need enormous calculations due to the complexities of their procedures. In this section we present a simple algorithm for scheduling multiple projects under multiple-resource constraints. The algorithm is an extension of TMROS with modifications to include multiple resources.

This algorithm assumes that the decision maker has priorities for the projects to be scheduled. According to these priorities we initially assign a specific number of each type of resources to each project. The algorithm then assigns resources to activities of the first project. There will be unused resources over different periods of time from the first project that will be added to the resource profile of the second project. Then the second project is scheduled according to the steps of the algorithm. Unused resources from the second project will be added to the resource profile of the third project and the scheduling process is repeated until all projects are scheduled.

The algorithm represents a control rule for calculating a criterion that includes all types of resources according to their importance to the network schedule. This control rule is equivalent to ACROS and it will be called

Table 6-14. COMPARISON BETWEEN TMROS, TG.2, AND THE OTHER TECHNIQUES

Network	Number of Nodes	Number of Activities	Maximum Resource Level	ACTIM	ACTRES	TIMRES	GENRES	ROT	ROT/ACTIM	ROT/ACTRES	TMROS	TG.2
1	5	7	3	20	20	20	20	17*	17*	17*	17*	17*
2	6	7	5	34	34	34	34	36	34	34	28*	28*
			6	34	34	34	34	28*	28*	28*	28*	28*
3	7	8	8	9	12	9	9	9	8*	9	8*	8*
			9	8*	12	9	8*	9	8*	9	8*	8*
4	7	9	6	60	58*	60	58*	60	60	58*	58*	58*
5	7	10	13	38	38	38	38	38	38	38	37*	37*
			14	34	34	34	34	37	34	34	33*	33*
6	8	11	5	77*	87	87	77*	87	77*	87	77*	77*
7	8	12	7	24	28	23*	23*	27	24	25	24	23*
			8	24	20	22	20	20	19*	19*	19*	19*
8	11	15	12	54	69	54	51*	61	51*	60	51*	51*
			13	51	64	64	51	52	50*	52	50*	50*
9	13	18	5	61	58	58	58	61	58	57*	58	57*
			6	53	52	52	52	59	50	49*	49*	49*
10	16	21	9	66	68	66	65*	75	66	68	66	65*
			10	57*	61	57*	57*	61	57*	57*	57*	57*
11	23	26	13	139	159	159	139	168	139	143	137*	137*
			14	130	131	144	130	136	130	131	121*	121*
			15	141	132	127	127	134	134	132	120*	120*
12	24	31	8	326	326	326	326	327	326	326	324*	324*
			9	311	318	311	311	311	304	311	303*	303*
13	22	35	9	125*	148	139	125*	143	125*	128	125*	125*
			10	116	122	122	116	118	115	117	110*	110*
14	20	38	13	130	121*	127	121*	137	125	121*	125	121*
			14	118	118	114*	114*	134	118	116	117	114*
			15	119	117	120	117	128	117	117	117	116*
15	33	48	7	303	293	286	285	295	283*	283*	283*	283*
			8	227	232	235	223	236	223	221	222	217*
			9	199	202	201	197	214	196	198	194*	194*

*Best project schedule.

210

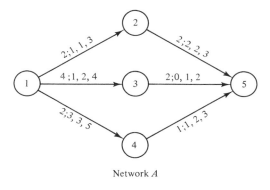

Network *A*

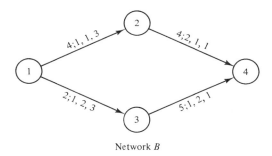

Network *B*

Figure 6-17. Multiple projects, multiple resources for Example 6-10. Time; resource 1, resource 2, resource 3.

eq.ACROS for convenience. The procedure of the algorithm is illustrated by the solution of a two-project, three-resource problem given in Example 6-10.

EXAMPLE 6-10

Schedule the two-project, three-resource problem shown in Fig. 6-17. Priorities of projects *A* and *B* are 60% and 40%, respectively. The numbers of resources available of types 1, 2, and 3 are 5, 8, and 11, respectively.

SOLUTION

The scheduling procedure is as follows:

1. Assign priorities to the projects to be scheduled. If no priorities are assigned, projects will be given equal priority values. In this example, the priority of project *A* is 60% and 40% for project *B*.
2. According to the priority of each project and the total number of resources available of each type, we initially assign a specific number of each type of

resource to each project. For example, assignments of resources to project A are

$$\text{resource } 1 \equiv \; 5 \times 0.6 = 3$$
$$\text{resource } 2 \equiv \; 8 \times 0.6 = 5 \text{ (rounded to higher integer)}$$
$$\text{resource } 3 \equiv 11 \times 0.6 = 7 \text{ (rounded to higher integer)}$$

Assignments of resources to project B are

$$\text{resource } 1 \equiv \; 5 - 3 = 2$$
$$\text{resource } 2 \equiv \; 8 - 5 = 3$$
$$\text{resource } 3 \equiv 11 - 7 = 4$$

This process becomes complicated when several types of resources and projects are considered. The following rules can be used to simplify the process:
(a) For the first project, if the number of any resource is real, it should be approximated to its higher integer value.
(b) The assignment of resources to the last project is accomplished by subtracting the total number of resources already assigned to other projects from the total number of resources available. This should be performed for each type of resources.
3. For each project, test the resource assignments for each activity against the available resources for the project to check for feasible schedules. The schedule is infeasible if any activity requires more resources than the total number of available resources of this type. If the schedule is infeasible, change the profile of resources to achieve feasibility of the schedules.
4. Calculate ACTIM for each activity in project A. The equivalent ACROS values (eq.ACROS) for the activities in project A are determined as follows:
(a) Calculate the total number of each type of resources needed by all activities in the network on all paths. In this example, 8, 11, and 20 resources are needed for resources type 1, 2, and 3, respectively.
(b) For each type of resources calculate (A_r/S_r) value where

A_r = minimum number of resources available of type r for this project
S_r = total number of resource r in all paths of the network

For this example,

$$\text{Resource } 1: \frac{A_1}{S_1} = \frac{3}{8} = 0.38$$

$$\text{Resource } 2: \frac{A_2}{S_2} = \frac{5}{11} = 0.45$$

$$\text{Resource } 3: \frac{A_3}{S_3} = \frac{7}{20} = 0.35$$

(c) Divide each type of resources needed for an activity by its corresponding value of A_r/S_r and sum the resulting values. This sum is referred to as the equivalent resource value of the activity. The resource values of all activities of project A are

$$\text{activity } 1\text{-}2 = \frac{1}{0.38} + \frac{1}{0.45} + \frac{3}{0.35} = 13.43$$

$$\text{activity } 1\text{-}3 = \frac{1}{0.38} + \frac{2}{0.45} + \frac{4}{0.35} = 18.50$$

$$\text{activity } 1\text{-}4 = 28.85$$

$$\text{activity } 2\text{-}5 = 18.28$$

$$\text{activity } 3\text{-}5 = 7.94$$

$$\text{activity } 4\text{-}5 = 15.65$$

(d) Now, eq.ACROS value of an activity is calculated as the maximum equivalent resource value it controls through the network on any one path.

Activity	eq.ACROS
1–2	31.71
1–3	26.44
1–4	44.50
2–5	18.28
3–5	7.94
4–5	15.65

5. Resources are assigned to activities of project A, according to a combined weights of eq.ACROS and ACTIM as given in TMROS. Table 6-15 shows eq.ACROS and ACTIM values for the activities of project A. The criterion, AG3, which combines eq.ACROS and ACTIM, is calculated as follows:

$$AG3 = W(\text{eq.ACROS}) + (1 - W)(\text{ACTIM})$$

Table 6-15. ACTIM AND eq.ACROS FOR ACTIVITIES OF PROJECT A

	eq.ACROS		ACTIM	
Activity	Value	0–100 Scale	Value	0–100 Scale
1–2	31.71	71.26	4	66.67
1–3	26.44	59.42	6	100.00
1–4	44.50	100.00	3	50.00
2–5	18.28	41.08	2	33.33
3–5	7.94	17.84	2	33.33
4–5	15.65	35.17	1	16.67

Table 6-16. SCHEDULE OF PROJECT A AT $W = 0.40$

Sequence of activities: 1–3, 1–4, 1–2, 2–5, 3–5, 4–5.

TNOW	Activity	Duration	TSTART	TFIN	Type of Resource Available 1	2	3	Activities Allowed
0	—	—	—	—	3	5	7	1–3, 1–4, 1–2
0	1–3	4	0	4	2	3	3	
0	1–2	2	0	2	1	2	0	
2	—	—	—	—	2	3	3	1–4, 2–5
2	2–5	2	2	4	0	1	0	
4	—	—	—	—	3	5	7	1–4, 3–5
4	1–4	2	4	6	0	2	2	
4	3–5	2	4	6	0	1	0	
6	—	—	—	—	3	5	7	4–5
6	4–5	1	6	7	2	3	4	

6. Estimate the project duration for different weights, W. The best schedule is the one with the least project duration. Tables 6-16 and 6-17 present the schedules for project A at $W = 0.4$ and 0.7, respectively. The project durations for $W = 0.4$ is 7 and 8 at $W = 0.7$.

7. Construct the actual profile of resources for project A. Then add the unused resources in the first project (project A) to the resource profile of the second project. The actual resources profile of project A and the modified resources profile of project B are shown in Fig. 6-18.

8. Now schedule the second project (project B) according to its modified resources profile. The steps above are repeated for other projects. Tables 6-18 and 6-19 show the project schedules of project B with $W = 0.5$ and 0.9, respectively. It should be noted that the minimum completion time of the project is 10 units of time.

Table 6-17. SCHEDULE OF PROJECT A AT $W = 0.7$

Sequence of activities: 1–4, 1–3, 1–2, 2–5, 4–5, 3–5.

TNOW	Activity	Duration	TSTART	TFIN	Type of Resource Available 1	2	3	Activities Allowed
0	—	—	—	—	3	5	7	1–4, 1–3, 1–2
0	1–4	2	0	2	0	2	2	
2	—	—	—	—	3	5	7	1–3, 1–2, 4–5
2	1–3	4	2	6	2	3	3	
2	1–2	2	2	4	1	2	0	
4	—	—	—	—	2	3	3	2–5, 4–5
4	2–5	2	4	6	0	1	0	
6	—	—	—	—	3	5	7	4–5, 3–5
6	4–5	1	6	7	2	3	4	
6	3–5	2	6	8	2	2	2	

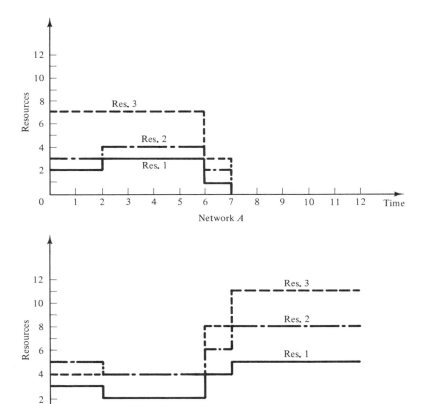

Figure 6-18. Resource profiles of projects *A* and *B*.

Table 6-18. SCHEDULE OF PROJECT *B* AT *W* = 0.5

Sequence of Activities: 1–2, 1–3, 3–4, 2–4.

TNOW	Activity	Duration	TSTART	TFIN	Type of Resource Available 1	2	3	Activities Allowed
0	—	—	—	—	3	4	5	1–2, 1–3
0	1–2	4	0	4	2	3	2	
4	—	—	—	—	2	4	4	1–3, 2–4
4	1–3	2	4	6	1	2	1	
6	—	—	—	—	4	6	8	3–4, 2–4
6	3–4	5	6	11	3	4	7	
11	—	—	—	—	2	4	4	2–4
11	2–4	4	11	15	0	3	3	

Table 6-19. SCHEDULE OF PROJECT B AT $W = 0.9$

Sequence of Activities: 1–3, 1–2, 3–4, 2–4.

TNOW	Activity	Duration	TSTART	TFIN	Type of Resource Available 1	2	3	Activities Allowed
0	—	—	—	—	3	5	4	1–3, 1–2
0	1–3	2	0	2	2	3	1	
2	—	—	—	—	2	4	4	1–2, 3–4
2	1–2	4	2	6	1	3	1	
2	3–4	5	2	7	0	1	0	
6	—	—	—	—	3	4	7	2–4
6	2–4	4	6	10	1	3	6	

PROBLEMS

6.1. Construct a project network with the following precedence relationships:

$$A < C$$
$$B < D$$
$$C < E, F$$
$$D < E, F$$
$$E < G$$
$$F < H$$
$$G < I$$
$$H < J$$
$$I < K$$
$$J < K$$

Durations of these activities are:

Activity	Duration
A	10
B	20
C	7
D	3
E	5
F	15
G	3
H	4
I	10
J	5
K	6

(a) Construct a network to represent the activities listed above. Check the network for consistency.

(b) Use linear programming to determine the critical path of the network.

6.2. Construct a project network with the following precedence relationships:

$$A < B, C$$
$$B < D$$
$$C < E, D$$
$$D < F$$
$$E < F$$

The durations of the activities are as follows:

A	10
B	7
C	12
D	5
E	4
F	6

(a) Find the critical path of this project.

(b) If the duration of activity D is reduced to 2 days, what is the effect on the critical path?

6.3. The following is a set of activities required in assembling a small model of an automated storage/automated retrieval system:

Activity	Description	Immediate Predecessor	Normal Duration (days)	Normal Cost ($)	Crash Duration (days)	Crash Cost ($)	CC UT
A	Construct storage racks	—	10	100	5	150	10 \$/DAY
B	Assemble bins	—	8	120	6	200	+o
C	Construct base	A	5	200	4	220	20
D	Build tracks	—	4	80	3	90	10
E	Construct crane	D	12	300	10	400	50
F	Assemble storage container	E	5	100	5	100	∞
G	Install motors	E	7	150	5	200	25
H	Install gear trains	C, E	4	50	3	60	10
I	Connect computer controls	H, A, B	10	600	8	750	75
J	Test system	I	5	400	4	510	110

(a) Set up a project network and determine the earliest and latest start and finish times for all activities on a normal basis.

(b) What is the project cost?

(c) Suggest some activities to be substituted on a crash basis and compute the new project cost.

6.4. *Manufacturing and test of a flight propulsion system:* In support of the U.S. military forces worldwide, the Defense Meteorological Satellite Program (DMSP) utilizes a sophisticated operational system of spacecraft and ground receiving stations to disseminate timely, high-quality meteorological data. This global weather information is supplied by a weather satellite built by R.C.A. One of the key subsystems that is a part of the satellite is the flight propulsion system. The propulsion subsystem, which is integrated as a complete package, consists of propellant tanks, rocket engines, and other components. After all components such as rocket engines and propellant tanks are on hand, they are assembled into a cylindrical structure which is a riveted, reinforced-aluminum alloy shell. Propellant tubing is then formed and brazed together into propellant manifolds. These manifolds are then connected to the tanks, rocket engines, and other components. The manufacturing of flight hardware for such a system is a complex task that requires frequent inspections and testing. A project that encompasses such fabrication and testing is manpower intensive and must meet strict schedule requirements. A list of the main activities of this project is given below. What is the probability of completing the project in 42 weeks, 52 weeks, and 46 weeks?

ACTIVITY LIST

Activity Code	Activity Description	Activity Dependence
A	Accumulate hydrazine lines	—
B	Input	A
C	Orifice calculations	B
D	Orifice tank outlets	C
E	Orifice bypass lines	D
F	Lines cleaned	E
G	Inspect hydrazine lines	F
H	Layout tube runs	—
I	Tube design	H
J	Band tubing	I
K	Cut and clean tubing	J
L	Brazing	K
M	Brazing	K
N	Brazing	K
O	Brazing	K
P	X-ray	L
Q	X-ray	M
R	X-ray	N
S	X-ray	O
T	Final inspection	P, Q, R, S
U	Prepare structure	—
V	Paint and tape	U
W	Inspect	V
X	Add tank supports	W
Y	Accumulate tanks	—
Z	Mount tanks	X, Y

ACTIVITY LIST (CONTINUED)

Activity Code	Activity Description	Activity Dependence
AA	Install regulator	Z
BB	Install fill and drain valves	AA
CC	Braze regulator	BB
DD	Manifold installation	T, CC
EE	Accumulate engines	—
FF	NEA'S modified	EE
GG	NEA'S cleaned	FF
HH	Braze unions	GG
II	X-ray	HH
JJ	Braze filters—NEAS	II
KK	Install thermostats	JJ
LL	Test thermostats	KK
MM	Install blankets	LL
NN	Install brackets	MM
OO	Complete NEA assembly	NN
PP	Braze tanks	DD
QQ	X-ray	OO, PP
RR	LVA brazed	QQ
SS	Install hydrazine lines	RR
TT	Install NEAS	G, SS
UU	Acceptance testing	TT
VV	RCA/DCAS final buy-off	UU

NODE AND DURATION LIST

Activity Code	Node I	Node J	Activity Duration t_a	t_m	t_b
A	1	2	0	1	2
B	2	3	1	1	1
C	3	4	1	2	3
D	4	5	1	1	1
E	5	6	2	3	4
F	6	7	1	2	3
G	7	39	1	1	1
H	1	8	5	10	15
I	8	9	1	2	3
J	9	10	2	3	4
K	10	11	6	8	10
L	11	12	5	10	15
M	11	13	4	6	8
N	11	14	2	4	6
O	11	15	2	3	4
P	12	16	2	3	4
Q	13	16	2	3	4
R	14	16	1	2	3
S	15	16	1	2	3

NODE AND DURATION LIST (CONTINUED)

Activity Code	Node I	Node J	Activity Duration t_a	t_m	t_b
T	16	24	1	1	1
U	1	17	1	2	3
V	17	18	1	2	3
W	18	19	1	2	3
X	19	20	2	5	8
Y	1	20	1	1	1
Z	20	21	2	3	4
AA	21	22	1	1	1
BB	22	23	1	1	1
CC	23	24	1	2	3
DD	24	25	2	4	6
EE	1	26	5	10	15
FF	26	27	1	2	3
GG	27	28	1	2	3
HH	28	29	1	1	1
II	29	30	1	2	3
JJ	30	31	1	2	3
KK	31	32	3	5	7
LL	32	33	1	1	1
MM	33	34	1	1	1
NN	34	35	1	1	1
OO	35	36	2	4	6
PP	25	36	1	2	3
QQ	36	37	1	2	3
RR	37	38	1	2	3
SS	38	39	2	3	4
TT	39	40	3	5	7
UU	40	41	2	6	10
VV	41	42	3	4	5

6-5. Construct a network based on the following restriction list:

Activity	Duration (days) Optimistic	Most Likely	Pessimistic	Restriction
A	8	10	14	A < E, F
B	0.5	1	2	B < C, D, E, F
C	16	20	25	C < H
D	3	5	8	D < G, H
E	7	10	12	E < G, H
F	0.5	2	4	
G	8	12	17	
H	0.75	1	1.5	

(a) Check the consistency of your network.

P189 →(b) Determine the critical path by the tabular method.

P198 →(c) Determine the probability of completing the project 3 days earlier. Interpret such probability.

(d) Determine the sequence of activities that constitute the second and third critical paths, that is, the paths with the second and third longest duration.

(e) Based on the first, second, and third CPs determined in part (d), determine the four most critical activities. Explain your reasons for choosing these four activities.

(f) Find a time, say x, such that the probability of project completion time exceeding x is less than approximately 0.10.

(g) If the available resources for this network are 3 units and the resources required for activities A through H are:

Activity	Resources Required
A	2
B	1
C	2
D	1
E	1
F	2
G	1
H	1

(1) Find the duration to complete all activities using the ACTIM approach.

(2) Find the duration to complete all activities using ACTRES, ROT, and TIMRES.

6-6. (a) Find the crashing schedule that minimizes the cost of the project network shown in Fig. 6-19.

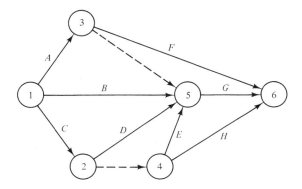

Figure 6-19. Network for Prob. 6-6.

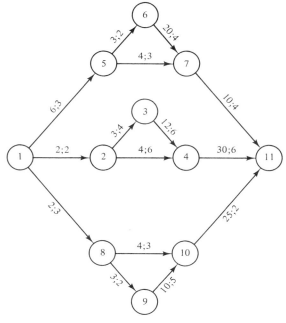

Figure 6-20. Network for Prob. 6-7. Duration; resource.

	Time (days)		Cost ($)	
Activity	Normal	Crash	Normal	Crash
A	10	8	180	200
B	12	11	400	450
C	7	7	500	500
D	4	3	160	200
E	8	6	80	100
F	12	11	500	600
G	9	9	350	350
H	16	13	610	700

(b) Using the normal times of the activities and the following resource requirements, determine the minimum project duration according to TIMRES.

Activity	Resources Required
A	3
B	2
C	1
D	4
E	2
F	3
G	2
H	3

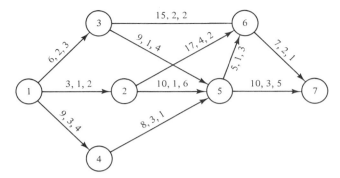

Figure 6-21. Network for Prob. 6-8. Duration; resource 1, resource 2.

Resources available are 5 units.

6-7. Given the project network shown in Fig. 6-20, with arrows as activities and the numbers by each arrow representing job durations in days and the number of resources required for the activity, assuming that 8 units of resources are available, find the minimum project completion time using ROT, TMROS (for $W = 0.5$), and ACTRES. Develop your own criterion for resource scheduling and compare results.

6-8. Use the AG.3 criterion to find the minimum completion time of the multiple-resource project given in Fig. 6-21. The maximum number of resources of types 1 and 2 are 6 and 8, respectively.

REFERENCES

[1] BATTERSBY, ALBERT, *Network Analysis for Planning and Scheduling.* New York: St. Martin's Press, Inc., 1970.

[2] BEDWORTH, DAVID D., *Industrial Systems: Planning, Analysis, Control.* New York: The Ronald Press Company, 1973.

[3] BELLMAN, RICHARD E., AND S. E. DREYFUS, *Applied Dynamic Programming.* Princeton, N.J.: Princeton University Press, 1962.

[4] BROOKS, GEORGE H., *Algorithm for Activity Scheduling to Satisfy Constraints on Resource Availability,* unpublished paper, September 17, 1963.

[5] DAVIS, E. W., "CPM Use in Top 400 Construction Firms," *Journal of the Construction Division, ASCE, 100* (March 1974).

[6] ELSAYED, ELSAYED A., "Algorithm for Project Scheduling with Resource Constraints," *International Journal of Production Research, 20,* no. 1 (1982).

[7] ESOGBUE, AUGUSTINE O., AND B. R. MARKS, "Dynamic Programming Models of the Non-serial Critical Path-Cost Problem," *Management Science, 24,* no. 2 (October 1977).

[8] GAITHER, NORMAN, "The Adoption of Operations Research Techniques by Manufacturing Organizations," *Journal of Decision Sciences, 6*, no. 4 (October 1975).

[9] HARRIS, R. G., *Precedence and Arrow Network Techniques for Construction.* New York: John Wiley & Sons, Inc., 1978.

[10] HOROWITZ, JOSEPH, *Critical Path Scheduling.* New York: The Ronald Press Company, 1967.

[11] KELLEY, J. E. JR., AND M. R. WALKER, "Critical Path Planning and Scheduling," *Proceedings of Eastern Joint Computer Conference*, Boston, 1959.

[12] KRAKOWSKI, MARTIN, "PERT and Parkinsin's Law," *Interfaces, 5*, no. 1 (November 1974).

[13] LEVIN, RICHARD I., AND C. A. KIRKPATRICK, *Planning and Control with PERT-CPM.* New York: McGraw-Hill Book Company, 1966.

[14] LOFTS, N. R., "Multiple Allocation of Resources in a Network—An Optimal Scheduling Algorithm," *Information, 1*, no. 12 (1974).

[15] MACCRIMMON, KENNETH R., AND CHARLES A. RYAVEC, "Analytical Studies of the PERT Assumptions," *Operations Research, 12*, no. 1 (January–February 1964).

[16] MARIMONT, ROSALIND B., "A New Method of Checking the Consistency of Precedence Matrices," *JACM, 6*, no. 2 (1959).

[17] MASON, RICHARD, "An Adaptation of the Brook's Algorithm for Scheduling Projects Under Multiple Source Constraints," unpublished M.S. thesis, Arizona State University, 1970.

[18] MILLER, L. C., *Successful Management for Contractors,* New York: McGraw-Hill Book Company, 1962.

[19] MODER, JOSEPH J., AND CECIL R. PHILLIPS, *Project Management with CPM and PERT.* New York: Reinhold Publishing Corporation, 1964.

[20] MOORE, LAWRENCE J., AND EDWARD R. CLAYTON, GERT *Modeling and Simulation: Fundamentals and Applications.* New York: Petrocelli-Charter, 1976.

[21] NASR, NABIL Z., "New Algorithms for Project Scheduling Under Resource Constraints," unpublished M.S. thesis, Rutgers University, 1983.

[22] PAIGE, HILLIARD W., "How PERT Cost Helps the General Manager," *Harvard Business Review*, November–December 1963.

[23] "PERT, Program Evaluation Research Task, Phase I Summary Report," *Bureau of Naval Weapons*, Dept. of the Navy, Washington, D. C., July, 1958.

[24] PETROVIC, RADIVOJ, "Optimization of Resource Allocation in Project Planning," *Operations Research, 3*, no. 16 (1968).

[25] PRITSKER, ALAN B., *Modeling and Analysis Using Q-GERT Networks.* New York: John Wiley & Sons, Inc., 1977.

[26] REIMHERR, G. W., *PERT (A Bibliography with Abstracts)*, National Technical Information Service, Final Report for 1964–1978, Springfield, Va., 1978.

[27] SAULS, E., "The Use of GERT," *Journal of Systems Management, 23*, no. 8 (1972), pp. 18–21.

[28] SCOTT, WILLIAM, "An Evaluation of Modifications and Extensions to Project Network Models," unpublished M.S. thesis, Rutgers University, 1979.

[29] WHITEHOUSE, GARY E., AND BEN L. WECHSLER, *Applied Operations Research.* New York: John Wiley & Sons, Inc., 1976.

[30] WHITEHOUSE, GARY E., AND JAMES R. BROWN, "GENRES: An Extension of Brooks Algorithm for Project Scheduling with Resource Constraints," *Computers and Industrial Engineering, 3* (1979).

[31] WIEST, JEROME D., AND FERDINAND K. LEVY, *Management Guide to PERT/CPM,* 2nd ed. Englewood Cliffs, N.J.: Prentice-Hall, Inc., 1969.

JOB SEQUENCING
AND OPERATIONS SCHEDULING

7.1. Introduction

This chapter deals with the most detailed analysis in production systems control: job sequencing and operations scheduling. The chapter is divided into two main sections: (1) job sequencing, in which machines process jobs such that some measure of performance is optimized; and (2) assembly line balancing and transfer lines, where assembly operations are assigned to several workstations along an assembly line in order to achieve equal balance between stations and increase the overall efficiency of the assembly line.

7.2. Job Sequencing

The job sequencing problem is considered to be one of the most interesting problems in production analysis. It has received the considerable attention of researchers. The problem is quite complex and far from being completely solved. Optimal solutions could be found for job sequencing problems with a small number of machines. However, optimal solutions for problems with a larger number of machines do not exist. In fact, it is impossible to check for optimality for such problems.

The job sequencing problem can be stated as follows:

Given n jobs to be processed, each has a setup time, processing time, and a due date. In order to be completed, each job is to be processed at several machines. It

7

is required to sequence these jobs on the machines in order to optimize a certain performance criterion.

A typical list of performance criteria to be optimized is:

1. Mean flow time, or mean time in the shop
2. Idle time of machines
3. Mean lateness of jobs (lateness of a job is defined as the difference between the actual completion time of the job and its due date)
4. Mean earliness of jobs (if a job is completed before its due date, then its lateness value is negative and it is referred to as earliness instead)
5. Mean tardiness of jobs (if a job is completed after its due date, then its lateness value is positive, and it is referred to as tardiness instead)
6. Mean queue time
7. Mean number of jobs in the system
8. Percentage of jobs late

The following factors serve to describe and classify a specific scheduling problem:

1. The number of jobs to be scheduled
2. The number of machines in the machine shop
3. Type of manufacturing facility (flow shop or job shop)

4. Manner in which jobs arrive at the facility (static or dynamic)
5. Criterion by which scheduling alternatives will be evaluated

The first factor defines the exact number of jobs to be processed, time required for each process, and the type of machine needed. The second factor defines the number of machines in the workshop. Factor 3 describes the flow of jobs through the workshop: if the flow is continuous and the jobs require the same sequence of machines, we have a *flow shop pattern*. In situations where there is no common pattern for the flow of jobs through the shop, a *job shop pattern* is said to exist. Dudek et al. [9] reported the results of an industrial survey which show that 57% of actual problems are of the job shop type, compared with 20% of the flow shop. They state that "a large portion of sequencing research deal with flow shop problems utilizing makespan as the objective criterion. On the other hand, most industrial problems fall in categories of job shop with or without dependence between jobs and the schedulers are faced with satisfying a multiple number of criteria."

The job arrival pattern is classified as either *static or dynamic*. In the *static* pattern, there are *n* jobs, each of which must be processed by a set of machines. All of the *n* jobs are available for processing at the initiation of the scheduling period, and no new jobs arrive during the period. In the *dynamic pattern*, jobs arrive intermittently according to a *stochastic process*.

The fifth factor describes the use of one or more criterion that are mentioned earlier in this section. The following example illustrates the factors discussed above.

EXAMPLE 7-1

Two jobs, *A* and *B*, are required to be scheduled on two machines, M1 and M2. Each job is processed first on M1, then on M2. The processing times of these jobs are:

		Processing Time	
		M1	*M2*
Jobs	A	8	5
	B	7	10

Find the sequence of these jobs that minimizes the makespan.

SOLUTION

number of jobs = 2   (factor 1)

number of machines = 2   (factor 2)

Each job is processed on M1 first, then on M2 [i.e., flow shop pattern (factor 3)]. The jobs are available at the beginning of the scheduling period [i.e., static job arrival

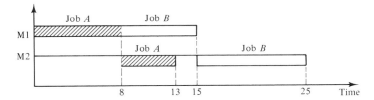

Figure 7-1. Job *A* is scheduled first.

pattern (factor 4)]. The criterion (factor 5) is to minimize the makespan schedule of the jobs (the total amount of time required to completely process all the jobs).

There are two possible sequences: we either start with job *A* or job *B*. Assuming that we start with job *A*, the total amount of time required to complete jobs *A* and *B* is 25 units of time, as shown in Fig. 7-1. If we start with job *B* first, the total time to complete jobs *A* and *B* is 22 units of time, as shown in Fig. 7-2.

If we increase the number of jobs (*n*) and the number of machines (*m*), the scheduling problem becomes more complicated. In fact, no optimal solutions exist for sequencing problems of large size *n* and *m*. Therefore, nearly all the reported studies have resorted to simulation and heuristic algorithms as the research tools.

We now summarize some techniques used for optimizing sequencing problems. The most obvious technique would be that of complete schedule enumeration, as we have shown in Example 7-1. This technique is useful for problems of a small number of jobs and machines, but becomes infeasible for a larger number of jobs and machines. For example, if there are *N* jobs, the number of possible sequences is *N*!.

The scheduling problem is further complicated in the case of the dynamic job arrival pattern. In this type of scheduling problem, a schedule for the current set of jobs is produced, and as this schedule is being worked through, new jobs arrive at the shop. The two extreme scheduling procedures are: to produce a new schedule each time a new job arrives, or to finish the existing schedule completely before producing a new schedule for the jobs that have arrived since the current schedule began [14]. Alternatively, a solution somewhere between these two extremes is a possibility. Ignoring computational considerations, the *best* approach may be to have an on-line system whereby

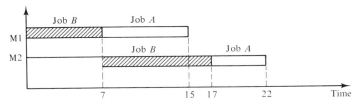

Figure 7-2. Job *B* is scheduled first.

every time an operation finishes on a machine and that information is fed into the computer, a new job is selected (based on a predetermined criterion) to be processed on that machine. The computational time is tremendous and in many cases it is not economically feasible. In this text we will limit our sequencing problems to those of static job arrival patterns.

If failure rates of the machines and the uncertainty of the processing time of jobs are to be considered, it becomes formidable, if not impossible, to solve this type of problems. Under these circumstances, simulation techniques are usually required.

Another technique to be used in solving the scheduling problem is the development of heuristic or search methods that will use rules that tend to give good, perhaps near optimal solutions. Unfortunately, with heuristics there is usually no way of checking how far from optimality the solution is. In presenting the sequencing problems, we first introduce the simplest one, followed by more complicated problems as we proceed.

7.3. n Jobs, One Machine

This is the simplest sequencing problem. There are n jobs to be processed by one machine; all jobs must go through this machine. Each job experiences two types of times: waiting time and processing time. Let W_i and t_i be the waiting time and processing time of job i, respectively. In order to sequence these jobs, a criterion for optimality must be defined. We assume that optimality is obtained if the mean flow time of the jobs is minimized. The *mean flow time* (MFT) is calculated as

$$MFT = \frac{\sum_{i=1}^{n} C_i}{n} \tag{7.1}$$

where MFT = mean flow time (average time in the shop)
C_i = completion time of job i ($C_i = W_i + t_i$)
n = number of jobs to be processed

EXAMPLE 7-2

Given below are the processing times of four jobs on machine X. Find the optimal sequence of jobs such that the mean flow time is minimized.

Job i	J1	J2	J3	J4
Processing time (t_i)	7	6	8	5

SOLUTION

Let us assume that the sequence $J1 \rightarrow J2 \rightarrow J3 \rightarrow J4$ is chosen (sequence A). Then

Job	W_i	t_i	C_i
J1	0	7	7
J2	7	6	13
J3	13	8	21
J4	21	5	26
	$\overline{41}$ +	$\overline{26}$ =	$\overline{67}$

$$\text{MFT}_{\text{sequence } A} = \tfrac{67}{4} = 16\tfrac{3}{4}$$

If sequence B, $J2 \rightarrow J3 \rightarrow J1 \rightarrow J4$, is chosen, then

Job	W_i	t_i	C_i
J2	0	6	6
J3	6	8	14
J1	14	7	21
J4	21	5	26
	$\overline{41}$ +	$\overline{26}$ =	$\overline{67}$

$$\text{MFT}_{\text{sequence } B} = \tfrac{67}{4} = 16\tfrac{3}{4}$$

There is no difference in MFT between sequences A and B.

Now we choose another sequence C such that the job with *shortest processing time* (SPT) is scheduled first. The sequence C is $J4 \rightarrow J2 \rightarrow J1 \rightarrow J3$. Therefore,

Job	W_i	t_i	C_i
J4	0	5	5
J2	5	6	11
J1	11	7	18
J3	18	8	26
	$\overline{34}$ +	$\overline{26}$ =	$\overline{60}$

$$\text{MFT}_{\text{sequence } C} = \tfrac{60}{4} = 15$$

For the single-machine case, the SPT sequencing rule guarantees a minimum mean flow time. (The reader is requested to prove this in Prob. 7-1.)

We conclude that if there are n jobs to be processed on one machine, the optimal sequence that minimizes the mean flow time should be based on the *shortest processing time* rule (SPT), where the job with the shortest processing time is scheduled first, followed by the job with the next shortest, and so on.

By employing the SPT rule, the mean job lateness is also minimized, where the lateness L_i of job i is defined as

$$L_i = C_i - d_i$$

where C_i = completion time of job i
$\quad d_i$ = due date of job i

The mean lateness (L) is defined as

$$L = \frac{1}{n} \sum_{i=1}^{n} L_i = \frac{1}{n} \sum_{i=1}^{n} (C_i - d_i)$$

$$= \frac{1}{n} \sum_{i=1}^{n} C_i - \frac{1}{n} \sum_{i=1}^{n} d_i = \text{MFT} - \bar{d}$$

where $\bar{d}$ is the average of the due dates of the n jobs and is therefore constant. Consequently, the minimization of L is achieved by using the SPT rule, which minimizes MFT.

If the sequencing criterion is to minimize the maximum lateness of n jobs, the optimum sequence is to process jobs in order of nondecreasing due dates, or

$$d_1 \le d_2 \le d_3 \cdots \le d_n$$

In some situations, we may assign certain weight g_i for job i to indicate the importance (priority) of the job (the larger g_i, the more important the job). The sequencing criterion in this case is to sequence the jobs in order to minimize the *mean weighted flow time* (MWFT), and the optimal schedule is obtained if the jobs are scheduled in the following manner:

$$\frac{t_1}{g_1} \le \frac{t_2}{g_2} \le \frac{t_3}{g_3} \le \cdots \le \frac{t_n}{g_n}$$

EXAMPLE 7-3

Assume that the weights assigned to jobs 1, 2, 3, and 4 in Example 7-2 are 7, 5, 10, and 3, respectively, on a scale of 0 to 10, where 10 represents the highest priority and 0 represents the least priority. Find the optimal sequence that minimizes the mean weighted flow time of the jobs.

SOLUTION

Job	t_i	g_i	t_i/g_i
J1	7	7	1.0
J2	6	5	1.2
J3	8	10	0.8
J4	5	3	1.66

The optimal sequence is J3 → J1 → J2 → J4.

7.4. n Jobs, Two Machines

Consider the situation where n jobs must be processed by two-machine centers M1 followed by M2. The processing times of all jobs on M1 and M2 are known and deterministic. It is required to find the optimal sequence that minimizes the makespan for the n jobs (i.e., the sequence that minimizes the time to complete all jobs). Johnson [19] developed an algorithm that can be used to obtain such an optimal sequence.

We introduce the bounds (limits) on the total processing times. The following two intuitive rules are considered:

1. The lower limit (L_1) on the total processing time may be obtained as

$$L_1 = \sum_{i=1}^{n} t_{i, \text{M1}} + t_{n, \text{M2}} \qquad (7.2)$$

where $t_{i, \text{M1}}$ = processing time of job i on machine 1
$t_{n, \text{M2}}$ = processing time of job n on machine 2

Equation (7.2) states that the last job (n) cannot start on M2 unless all jobs are processed on M1. Since $\sum_i t_{i, \text{M1}}$ is constant, this implies that it is desirable to place the job with the shortest processing time on machine 2 last in the sequence

2. The second intuitive rule for finding the lower limit (L_2) on the total processing time is

$$L_2 = t_{1, \text{M1}} + \sum_{i=1}^{n} t_{i, \text{M2}} \qquad (7.3)$$

which states that none of the jobs following the first job can be processed until the first job is processed on M1. Since $\sum_i t_{i, \text{M2}}$ is constant, this implies that it is desirable to place the job with the shortest processing time on machine 1 first in the sequence. The lower bound on processing all jobs is then obtained as

$$L = \max \{L_1, L_2\} \qquad (7.4)$$

The steps of Johnson's algorithm are summarized below:

1. List all processing times of all jobs on machines M1 and M2.
2. Scan through all processing times for all jobs. Locate the minimum processing time.
3. If the minimum processing time is on M1, place the corresponding job first (as early as possible) in the sequence. If it is on M2, place the corresponding job last (as late as possible) in the sequence.

4. Eliminate the assigned jobs (already placed in the sequence as a result of step 3) and repeat steps 2 and 3 until all jobs are sequenced.

5. A tie between two processing times is broken arbitrarily because it cannot affect the minimum elapsed time to complete all the jobs.

EXAMPLE 7-4

The drilling and riveting times for six jobs are given below. For every job, a hole is drilled first followed by riveting. Find the optimum sequence that minimizes the make-span for all jobs.

Job	J1	J2	J3	J4	J5	J6
Drilling	4	7	3	12	11	9
Riveting	11	7	10	8	10	13

SOLUTION

We refer to the drilling and riveting machines as M1 and M2, respectively. We also construct the following job sequence table. It has six elements (number of jobs).

Sequence Table

J3					

Scanning the listed times for all jobs on M1 and M2, we find that the shortest processing time is 3 for J3. Since it occurs on M1, J3 is sequenced as early as possible. We then delete J3 from the job listing:

Job	J1	J2	J4	J5	J6
M1	4	7	12	11	9
M2	11	7	8	10	13

Repeat step 2 of Johnson's algorithm. The shortest processing time is 4 for J1. We sequence J1 as early as possible; that is, it follows J3 in the sequence table as given below.

Sequence Table

J3	J1				

J1 is deleted from the job listing table.

Job	J2	J4	J5	J6
M1	7	12	11	9
M2	7	8	10	13

A tie exists between job times for J2 on M1 and M2. A decision to schedule J2 as late as possible (position 6) or as early as possible (position 3) in the sequence table is dependent on the priority (if there is any) of the job. If there is no priority, J2 can be placed in the third or the six position without affecting the makespan for all jobs. We now have two possible sequences.

Sequence Table

J3	J1				J2

Sequence Table

J3	J1	J2			

Eliminate J2 from the job listing.

Job	J4	J5	J6
M1	12	11	9
M2	8	10	13

The shortest processing time is 8, which corresponds to J4 on M2. Sequence J4 as late as possible.

Sequence Table

J3	J1			J4	J2

Sequence Table

J3	J1	J2			J4

J4 is eliminated from the job listing table.

Job	J5	J6
M1	11	9
M2	10	13

The shortest time of the unsequenced jobs is 9, which corresponds to J6 on M1. Therefore, we sequence J6 as early as possible and J5 is placed in the open position to complete the sequence. Either of the two sequences given below will minimize the makespan for all jobs.

Sequence Table

J3	J1	J6	J5	J4	J2

Sequence Table

J3	J1	J2	J6	J5	J4

7.5. *n* Jobs, Three Machines

This problem is an extension of *n* jobs, two machines. *Optimal solutions* can be found for this problem, where every solution requires that all jobs be processed in the same order on each machine. An optimal solution for the *n*-job, three-machine problem is obtained by:

1. Using Johnson's two-machine algorithm if certain conditions are met, or
2. Using Ignall and Schrage's *branch-and-bound algorithm*

7.5.1. JOHNSON'S ALGORITHM FOR *n* JOBS, THREE MACHINES

We consider three machines with the technical ordering M1, M2, and M3. No passing is allowed; that is, jobs must be processed in the same order on each machine. We define t_{ij} to be the processing time of job *i* on machine *j*. Johnson's algorithm for *n* jobs, two machines can be applied to *n* jobs, three machines and an optimal solution is obtained if either of the following conditions holds:

$$\min t_{i1} \geq \max t_{i2} \qquad (7.5)$$

or

$$\min t_{i3} \geq \max t_{i2} \qquad (7.6)$$

In other words, if the minimum processing time of all jobs on either machine 1 or machine 3 is greater than or equal to the maximum processing time of all jobs on machine 2, the three-machine Johnson's algorithm applies. We reformulate the problem by constructing two dummy machines (M'_1, M'_2) to replace the three existing machines. The processing time of job *i* on M'_1 is $t_{i1} + t_{i2}$ and the processing time on M'_2 is $t_{i2} + t_{i3}$. We then apply Johnson's algorithm for the two machines, M'_1 and M'_2, to find the optimal job sequence.

EXAMPLE 7-5

Find the optimal sequence for the following six jobs on M1, M2, and M3.

Job	Processing Times On:		
	M1	*M2*	*M3*
1	5	3	9
2	7	2	5
3	4	3	7
4	8	4	3
5	6	2	2
6	7	0	8

SOLUTION

We check for the conditions:

$$\min t_{i1} \geq \max t_{i2} \qquad i = 1, 2, \ldots, 6$$

or

$$\min t_{i3} \geq \max t_{i2} \qquad i = 1, 2, \ldots, 6$$

$$\min t_{i1} = \min \{t_{11}, t_{21}, t_{31}, t_{41}, t_{51}, t_{61}\} = 4$$

$$\max t_{i2} = \max \{t_{12}, t_{22}, t_{32}, t_{42}, t_{52}, t_{62}\} = 4$$

Since the condition in Eq. (7.5) is satisfied, we can apply Johnson's algorithm.

Job	M'_1	M'_2
1	8	12
2	9	7
3	7	10
4	12	7
5	8	4
6	7	8

Upon applying Johnson's algorithm, it is found that the optimal job sequences are:

$$3 \to 6 \to 1 \to 2 \to 4 \to 5, \quad 6 \to 3 \to 1 \to 2 \to 4 \to 5$$
$$3 \to 6 \to 1 \to 4 \to 2 \to 5, \quad 6 \to 3 \to 1 \to 4 \to 2 \to 5$$

Giglio and Wagner [12] tested Johnson's algorithm for a series of problems where the previously mentioned conditons [Eqs. (7.5) and (7.6)] did not hold. The algorithm was highly successful. In their test of 20 cases, the average time span for Johnson's schedule was 131.7 as compared to 127.9 for the optimal schedules. The optimal solutions were obtained for 9 out of the 20 cases. In 8 of the remaining cases, the optimal solutions could be obtained by making a single interchange in the ordering of two adjacent jobs.

In general, Johnson's approximation seems to yield a relatively large number of optimal schedules and a significantly large number of schedules that are nearly optimal.

7.5.2. BRANCH-AND-BOUND ALGORITHM

The branch-and-bound algorithm can be used when the conditions in Eqs. (7.5) and (7.6) are not satisfied and an optimal sequence must be determined. Ignall and Schrage [17] developed a branch-and-bound algorithm for the general three-machine flow shop problem. In this algorithm, the problem is represented by a tree structure where each node is a partial sequence. To determine which partial sequence we should branch from, we compute the lower bound on the makespan for all nodes, that is, the lower bounds on all partial sequences. We then branch from the node with the least lower bound. The process is continued until the sequence with the least lower bound is found. The lower bound on the makespan for all nodes is estimated as follows:

Consider n jobs $(1, 2, \ldots, n)$, and each job is to be processed on three machines M1, M2, and M3 in their respective order. The lower bound on the makespan for all jobs can be one of the following:

$$(1) \qquad L_1 = \sum_{i=1}^{n} t_{i, \text{M1}} + t_{n, \text{M2}} + t_{n, \text{M3}} \qquad (7.7)$$

In other words, the last job cannot start on M3 until it is completed on M2, and it cannot start on M2 until all jobs are completed on M1.

$$(2) \qquad L_2 = t_{1, \text{M1}} + \sum_{i=1}^{n} t_{i, \text{M2}} + t_{n, \text{M3}} \qquad (7.8)$$

Equation (7.8) states that job n cannot start on M3 until all jobs are completed on M2. Also, none of the jobs can start on M2 until job 1 is completed on M1:

$$(3) \qquad L_3 = t_{1, \text{M1}} + t_{1, \text{M2}} + \sum_{i=1}^{n} t_{i, \text{M3}} \qquad (7.9)$$

which implies that none of the jobs can be processed on M3 until job 1 is completed on M1 and M2.

We redefine the lower bounds so as to distinguish between the assigned and the unassigned jobs in the current solution. Ignall and Schrage [17] developed a lower bound for the makespan of all nodes emanating from a given node. The lower bound is determined as follows:

Consider node P, in a branching tree corresponding to sequence J_r, where J_r contains a particular subset (of size r) of n jobs. We define TIMEM1(J_r), TIMEM2(J_r), and TIMEM3(J_r) as the times at which machines M1, M2, and M3, respectively, complete processing the last job in the sequence J_r. Then a lower bound on the makespan of all schedules that begin with sequence J_r is

$$LB(J_r) = \max \left\{ \begin{array}{l} TIMEM1(J_r) + \sum_{\bar{J}_r} t_{i1} + \min_{\bar{J}_r} (t_{i2} + t_{i3}) \\ TIMEM2(J_r) + \sum_{\bar{J}_r} t_{i2} + \min_{\bar{J}_r} (t_{i3}) \\ TIMEM3(J_r) + \sum_{\bar{J}_r} t_{i3} \end{array} \right\} \quad (7.10)$$

where t_{ij} = processing time of job i on machine j ($i = 1, 2, \ldots, n$ and $j = 1, 2, 3$)

$\bar{J}_r$ = set of $n - r$ jobs that have not been assigned a position in sequence J_r

$LB(P) = LB(J_r)$ = lower bound on the makespan for any node that emanates from node P

After finding the lower bound on the nodes, we branch from the node with the smallest lower bound. Create a new node for every job that has not yet been scheduled. This is done by attaching an unscheduled job to the end of the sequence of scheduled jobs. The lower bounds are then computed for every new node by using Eq. (7.10). This process is repeated until a complete schedule is found and has the smallest lower bound.

EXAMPLE 7-6

Consider a four-job, three-machine flow shop scheduling problem given below. Use the branch-and-bound algorithm to find the sequence that minimizes the makespan of all jobs.

Job	Machine 1	Machine 2	Machine 3
1	14	6	15
2	8	11	4
3	10	13	17
4	16	15	5

SOLUTION

We branch from node 0 as shown in Fig. 7-3. The lower bounds on nodes 1, 2, 3, and 4 are computed by using Eq. (7.10). To calculate these lower bounds we need

TIMEM1(1) = 14   TIMEM2(1) = 20   TIMEM3(1) = 35

TIMEM1(2) = 8   TIMEM2(2) = 19   TIMEM3(2) = 23

TIMEM1(3) = 10   TIMEM2(3) = 23   TIMEM3(3) = 40

TIMEM1(4) = 16   TIMEM2(4) = 31   TIMEM3(4) = 36

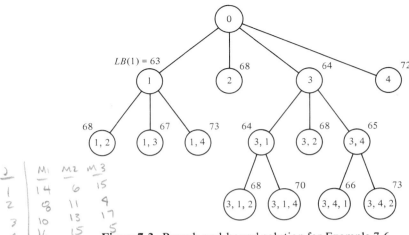

Figure 7-3. Branch-and-bound solution for Example 7-6.

The lower bounds are

$$LB(1) = \max \left\{ \begin{array}{l} 14 + 34 + 15 = 63 \\ 20 + 39 + 4 = 63 \\ 35 + 26 = 61 \end{array} \right\} = 63$$

$$LB(2) = \max \left\{ \begin{array}{l} 8 + 40 + 20 = 68 \\ 19 + 34 + 5 = 58 \\ 23 + 37 = 60 \end{array} \right\} = 68$$

$$LB(3) = \max \left\{ \begin{array}{l} 10 + 38 + 15 = 63 \\ 23 + 32 + 4 = 59 \\ 40 + 24 = 64 \end{array} \right\} = 64$$

$$LB(4) = \max \left\{ \begin{array}{l} 16 + 32 + 15 = 63 \\ 31 + 30 + 4 = 65 \\ 36 + 36 = 72 \end{array} \right\} = 72$$

Since LB(1) is the smallest lower bound, we branch from node 1. We need to calculate the times of sequence $1i$ ($i = 2, 3, 4$) as follows:

$$\text{TIMEM1}(12) = \text{TIMEM1}(1) + t_{21} = 14 + 8 = 22$$
$$\text{TIMEM2}(12) = \max \{\text{TIMEM1}(12) + t_{22}, \text{TIMEM2}(1) + t_{22}\}$$
$$= \max \{22 + 11, 20 + 11\} = 33$$

$$\text{TIMEM3}(12) = \max\{\text{TIMEM2}(12) + t_{23}, \text{TIMEM3}(1) + t_{23}\}$$
$$= \max\{33 + 4, 35 + 4\} = 39$$

$$\text{TIMEM1}(13) = \text{TIMEM1}(1) + t_{31} = 14 + 10 = 24$$

$$\text{TIMEM2}(13) = \max\{\text{TIMEM1}(13) + t_{32}, \text{TIMEM2}(1) + t_{32}\}$$
$$= \max\{24 + 13, 20 + 13\} = 37$$

$$\text{TIMEM3}(13) = \max\{\text{TIMEM2}(13) + t_{33}, \text{TIMEM3}(1) + t_{33}\}$$
$$= \max\{37 + 17, 35 + 17\} = 54$$

$$\text{TIMEM1}(14) = \text{TIMEM1}(1) + t_{41} = 14 + 16 = 30$$

$$\text{TIMEM2}(14) = \max\{\text{TIMEM1}(14) + t_{42}, \text{TIMEM2}(1) + t_{42}\}$$
$$= \max\{30 + 15, 20 + 15\} = 45$$

$$\text{TIMEM3}(14) = \max\{\text{TIMEM2}(14) + t_{43}, \text{TIMEM3}(1) + t_{43}\}$$
$$= \max\{45 + 5, 35 + 5\} = 50$$

The lower bounds are

$$\text{LB}(12) = \max \begin{cases} 22 + 26 + 20 = 68 \\ 33 + 28 + 5 = 66 \\ 39 + 22 = 61 \end{cases} = 68$$

$$\text{LB}(13) = \max \begin{cases} 24 + 24 + 15 = 63 \\ 37 + 26 + 4 = 67 \\ 54 + 9 = 63 \end{cases} = 67$$

$$\text{LB}(14) = \max \begin{cases} 30 + 18 + 15 = 63 \\ 45 + 24 + 4 = 73 \\ 50 + 21 = 71 \end{cases} = 73$$

Upon comparing all nodes at this stage, we find that $\text{LB}(3) = 64$ is the smallest lower bound; therefore, we branch from node 3.

$$\text{TIMEM1}(31) = \text{TIMEM1}(3) + t_{11} = 10 + 14 = 24$$

$$\text{TIMEM2}(31) = \max\{\text{TIMEM1}(31) + t_{12}, \text{TIMEM2}(3) + t_{12}\}$$
$$= \max\{24 + 6, 23 + 6\} = 30$$

$$\text{TIMEM3}(31) = \max\{\text{TIMEM2}(31) + t_{13}, \text{TIMEM3}(3) + t_{13}\}$$
$$= \max\{30 + 15, 40 + 15\} = 55$$

$$\text{TIMEM1}(32) = \text{TIMEM1}(3) + t_{21} = 10 + 8 = 18$$

$$\text{TIMEM2}(32) = \max \{\text{TIMEM1}(32) + t_{22}, \text{TIMEM2}(3) + t_{22}\}$$
$$= \max \{18 + 11, 23 + 11\} = 34$$
$$\text{TIMEM3}(32) = \max \{\text{TIMEM2}(32) + t_{23}, \text{TIMEM3}(3) + t_{23}\}$$
$$= \max \{34 + 4, 40 + 4\} = 44$$
$$\text{TIMEM1}(34) = \text{TIMEM1}(3) + t_{41} = 10 + 16 = 26$$
$$\text{TIMEM2}(34) = \max \{\text{TIMEM1}(34) + t_{42}, \text{TIMEM2}(3) + t_{42}\}$$
$$= \max \{26 + 15, 23 + 15\} = 41$$
$$\text{TIMEM3}(34) = \max \{\text{TIMEM2}(34) + t_{43}, \text{TIMEM3}(3) + t_{43}\}$$
$$= \max \{41 + 5, 40 + 5\} = 46$$

We now compute the lower bounds for the sequences 31, 32, and 34.

$$\text{LB}(31) = \max \left\{ \begin{array}{l} 24 + 24 + 15 = 63 \\ 30 + 26 + \ 4 = 60 \\ 55 + \ \ 9 \qquad = 64 \end{array} \right\} = 64$$

$$\text{LB}(32) = \max \left\{ \begin{array}{l} 18 + 30 + 20 = 68 \\ 34 + 21 + \ 5 = 60 \\ 44 + 20 \qquad = 64 \end{array} \right\} = 68$$

$$\text{LB}(34) = \max \left\{ \begin{array}{l} 26 + 22 + 15 = 63 \\ 41 + 17 + \ 4 = 62 \\ 46 + 19 \qquad = 65 \end{array} \right\} = 65$$

The smallest LB is LB(31) = 64. Therefore, we branch from node 31.

$$\text{TIMEM1}(312) = \text{TIMEM1}(31) + t_{21} = 24 + 8 = 32$$
$$\text{TIMEM2}(312) = \max \{\text{TIMEM1}(312) + t_{22}, \text{TIMEM2}(31) + t_{22}\}$$
$$= \max \{32 + 11, 30 + 11\} = 43$$
$$\text{TIMEM3}(312) = \max \{\text{TIMEM2}(312) + t_{23}, \text{TIMEM3}(31) + t_{23}\}$$
$$= \max \{43 + 4, 55 + 4\} = 59$$
$$\text{TIMEM1}(314) = \text{TIMEM1}(31) + t_{41} = 24 + 16 = 40$$
$$\text{TIMEM2}(314) = \max \{\text{TIMEM1}(314) + t_{42}, \text{TIMEM2}(31) + t_{42}\}$$
$$= \max \{40 + 15, 30 + 15\} = 55$$
$$\text{TIMEM3}(314) = \max \{\text{TIMEM2}(314) + t_{43}, \text{TIMEM3}(31) + t_{43}\}$$
$$= \max \{55 + 5, 55 + 5\} = 60$$

The lower bounds are

$$LB(312) = \max \begin{cases} 32 + 16 + 20 = 68 \\ 43 + 15 + 5 = 63 \\ 59 + 5 = 64 \end{cases} = 68$$

$$LB(314) = \max \begin{cases} 40 + 8 + 15 = 63 \\ 55 + 11 + 4 = 70 \\ 60 + 4 = 64 \end{cases} = 70$$

As LB(34) = 65 is the smallest LB, we branch from node (34) as follows:

TIMEM1(341) = TIMEM1(34) + t_{11} = 26 + 14 = 40

TIMEM2(341) = max {TIMEM1(341) + t_{12}, TIMEM2(34) + t_{12}}

 = max {40 + 6, 41 + 6} = 47

TIMEM3(341) = max {TIMEM2(341) + t_{13}, TIMEM3(34) + t_{13}}

 = max {47 + 15, 46 + 15} = 62

TIMEM1(342) = TIMEM1(34) + t_{21} = 26 + 8 = 34

TIMEM2(342) = max {TIMEM1(342) + t_{22}, TIMEM2(34) + t_{22}}

 = max {34 + 11, 41 + 11} = 52

TIMEM3(342) = max {TIMEM2(342) + t_{23}, TIMEM3(34) + t_{23}}

 = max {52 + 4, 46 + 4} = 56

The lower bounds are

$$LB(341) = \max \begin{cases} 40 + 8 + 15 = 63 \\ 47 + 11 + 4 = 62 \\ 62 + 4 = 66 \end{cases} = 66$$

$$LB(342) = \max \begin{cases} 34 + 14 + 21 = 69 \\ 52 + 6 + 15 = 73 \\ 56 + 15 = 71 \end{cases} = 73$$

The smallest lower bound on the tree, as shown in Fig. 7-3, is LB(341) = 66. We have an optimal sequence $3 \rightarrow 4 \rightarrow 1 \rightarrow 2$ with a makespan of 66 units of time.

7.6. Two Jobs, *M* Machines

This problem can be stated as follows: given two jobs *A* and *B*, each job has different processing sequences on *M* machines. It is required to determine the optimal sequences of these jobs on the *M* machines such that the total processing time is minimized. Beckman [4], Akers and Friedman [1], and Hardgrave and Nemhauser [14] proposed graphical models which yield optimum solutions to this problem. The solution approach is summarized below:

1. Construct a two-dimensional graph where the x axis represents the processing time and sequence of job 1 on the M machines, while y axis represents those of job 2. (Use the same scale for both x and y.)
2. Shade the areas where a machine would be occupied by the two jobs at the same time.
3. The processing of both jobs can be represented by a continuous path which consists of horizontal, vertical, and 45 degree diagonal segments. The path starts at the lower left corner and stops at the upper right corner, while avoiding the shaded areas in the graph. In other words, the path is not allowed to pass through shaded areas which correspond to operating both jobs concurrently on the same machine. Since a diagonal segment implies that both jobs are being processed on different machines at the same time, a feasible path that maximizes diagonal movement will minimize the total processing time. In other words, our aim is to have as much diagonal travel as possible. The schedule is determined by trial and error. Usually, only a few lines may be drawn before an optimal path is found.

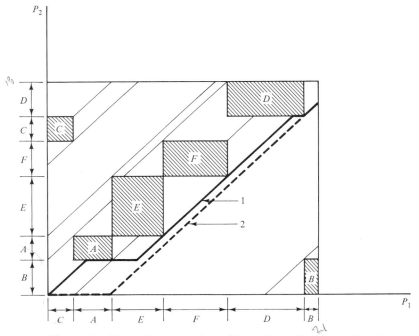

Figure 7-4. Processing two jobs on six machines for Example 7-7.

EXAMPLE 7-7

A product P passes through six machine centers during its manufacturing. There are two major parts for this product, P_1 and P_2. The technological sequence of these parts on the six machines and the manufacturing time on each machine are:

P_1: Machine sequence $C \rightarrow A \rightarrow E \rightarrow F \rightarrow D \rightarrow B$
    Time (hours)   2   3   4   5   6   1   21

P_2: Machine sequence $B \rightarrow A \rightarrow E \rightarrow F \rightarrow C \rightarrow D$
    Time (hours)   3   2   5   3   2   3   18

Find the optimal scheduling that minimizes the makespan of the jobs.

SOLUTION

We construct a two-dimensional graph where the horizontal axis represents P_1 and the vertical axis represents P_2. The shaded areas in Fig. 7-4 represent a time period during which the two jobs require the use of the same machine. Lines 1 and 2 maximize the diagonal travel from the bottom left corner to the upper right corner. The total makespan is ~~24~~ hours.
23

7.7. *n* Jobs, *M* Machines

This problem is a typical static flow shop sequencing problem where n jobs must be processed by a set of M machines. All jobs are processed at the initiation of the scheduling period, and no new jobs arrive during the period (static job arrival pattern). Also, jobs are not allowed to pass each other (i.e., jobs maintain the same position in the sequence). The problem is to schedule n jobs on M machines such that the jobs are completed in a minimum span of time. Unfortunately, there is no general solution for any problem where $M > 3$. However, there are some heuristic techniques that may obtain good sequences or may obtain optimal sequences (in many cases it is impossible to check for optimality). We present two heuristic algorithms that provide *good* schedules.

7.7.1. CAMPBELL ET AL. ALGORITHM

Campbell et al. [7] developed an algorithm that generates a series of sums for each job similar to the two sets of sums generated in the n-job, three-machine problem (M_1 and M_2). With M machines, $M - 1$ two column sets of job times can be developed and can then be solved using Johnson's algorithm for n jobs, two machines.

EXAMPLE 7-8

Find the *best* sequence for four jobs to be processed on four machines by using Campbell's algorithm. The processing times for each job are given as:

PROCESSING TIME

Job	Machine 1	Machine 2	Machine 3	Machine 4
1	5	6	4	5
2	4	7	3	5
3	9	5	5	3
4	6	8	4	1

SOLUTION

We develop the first set of machines as:

Job	M1	M4
1	5	5
2	4	5
3	9	3
4	6	1

Applying Johnson's algorithm, we obtain the optimal sequence $2 \rightarrow 1 \rightarrow 3 \rightarrow 4$. The second set of machines to be constructed is M12 and M34.

Job i	M12 $t_{i1} + t_{i2}$	M34 $t_{i3} + t_{i4}$
1	11	9
2	11	8
3	14	8
4	14	5

Again, we apply Johnson's algorithm to obtain the optimal sequence $1 \rightarrow 2 \rightarrow 3 \rightarrow 4$. The final set of machines is M123 and M234.

Job i	M123 $t_{i1} + t_{i2} + t_{i3}$	M234 $t_{i2} + t_{i3} + t_{i4}$
1	15	15
2	14	15
3	19	13
4	18	13

and the optimal sequence is $2 \rightarrow 1 \rightarrow 3 \rightarrow 4$.

From the results above, we find that there are two possible sequences: $1 \rightarrow 2 \rightarrow 3 \rightarrow 4$ and $2 \rightarrow 1 \rightarrow 3 \rightarrow 4$. These two sequences yield the same makespan of 37 units of time. In cases where many schedules are obtained by the algorithm, we should select the schedule that minimizes the makespan of all jobs.

7.7.2. STINSON–SMITH ALGORITHM

Stinson and Smith [30] developed a heuristic algorithm for sequencing the n-job, m-machine static flow shop problem. They approach the problem by first testing each of the n jobs as a potential immediate predecessor to each of the other jobs and a cost C_{ij} is determined for each job i if it were a predecessor to another job j. These values of C_{ij} are then used to produce a final schedule by solving the traveling salesman problem (see Section 7.8).

Let $q(j, k)$ be the completion time of job j on machine k and $t(j, k)$ be the time to process job j on some machine k. Consider only one pair of jobs, i and j, which are to be processed by the m machines, and job i precedes job j. The relationship between $q(i, k)$ and $q(j, k - 1)$ is one of the following:

1. $q(i, k) > q(j, k - 1)$; that is, job j will arrive at machine k prior to the time that job i will have released machine k; hence job j will be blocked in the queue at machine k for $q(i, k) - q(j, k - 1)$ time units.
2. $q(i, k) < q(j, k - 1)$; that is, job j will arrive at machine k after the time that job i has already released machine k; hence machine k will be idle for $q(j, k - 1) - q(i, k)$ time units.
3. $q(i, k) = q(j, k - 1)$; that is, job j will arrive at machine k at the instant where job i is released from machine k. In this case neither a block to job j nor idleness to machine k would result.

Consider the set of circumstances that would have to take place if $q(i, k) = q(j, k - 1)$ for the entire period where both i and j are jointly in process in the schedule. It is obvious that this will only occur when $t(i, k) = t(j, k - 1)$ for every machine in the range $2 \leq k \leq m$. Although this seldom happens, we still may recognize that the closer we can match $t(i, k)$ and $t(j, k - 1)$ values for all machines, the smoother jobs i and j will tend to fit together within the schedule. The purpose of the Stinson–Smith algorithm is to minimize the residual $r(ij, k)$, which is defined as

$$r(ij, k) = t(i, k) - t(j, k - 1) \qquad 2 \leq k \leq m \qquad (7.11)$$

For any pair of i and j and $i \neq j$, we may compute $m - 1$ such residuals. These residuals are then combined to yield an overall cost, C_{ij}, when job i precedes job j. Six different heuristic rules for combining these residuals are used as follows:

Heuristic 1—Sum of the absolute residuals (H1):

$$C_{ij} = \sum_{k=2}^{m} |r(ij, k)| \qquad (7.12)$$

Heuristic 2—Sum of the residuals squared (H2):

$$C_{ij} = \sum_{k=2}^{m} [r(ij, k)]^2 \tag{7.13}$$

The next four heuristics depend on the sign of each $r(ij, k)$ value. As indicated earlier, a positive $r(ij, k)$ indicates blocking of job j at machine k, while a negative $r(ij, k)$ indicates idleness of machine k. If it is more serious to incur machine idleness than delays of jobs at some machines, then heavier weights should be placed on the negative residuals. Further, it should also be recognized that a negative residual at some machine k has a carryover effect on other machines downstream of k. It could be proven that the carryover of negative residuals, $r^*(ij, k)$, may be generalized as follows:

$$r^*(ij, k) = t(i, k) - t(j, k - 1) + \min \{r(ij, k - 1); 0\} \tag{7.14}$$

Heuristics 3 through 6 are designed to either apply heavier weights on negative residuals or implement negative residual carryover or both.

Heuristic 3—Sum of only the negative residuals with negative residual carryover (H3):

$$C_{ij} = \sum_{k=2}^{m} |\min \{r^*(ij, k); 0\}| \tag{7.15}$$

Heuristic 4—Sum of absolute residuals with negative residual carryover (H4):

$$C_{ij} = \sum_{k=2}^{m} |r^*(ij, k)| \tag{7.16}$$

Heuristic 5—Sum of the residuals squared with negative residual carryover (H5):

$$C_{ij} = \sum_{k=2}^{m} [r^*(ij, k)]^2 \tag{7.17}$$

Heuristic 6—Sum of the absolute residuals with negative residuals weighted double—no carryover (H6):

$$C_{ij} = \sum_{k=2}^{m} \max \{r(ij, k); 0\} + 2 |\min \{r(ij, k); 0\}| \tag{7.18}$$

After calculating all C_{ij} for each pair of jobs i and j, a cost matrix is then constructed and a traveling salesman algorithm is applied to find the optimal sequence. We now introduce a numerical example to illustrate the use of the foregoing heuristics.

EXAMPLE 7-9

Find the best sequence for four jobs to be processed on three machines by using H1 (as stated in the Stinson–Smith algorithm). The processing times for each job on all machines are:

PROCESSING TIME

Job	Machine 1	Machine 2	Machine 3
1	14	6	15
2	8	11	4
3	10	13	17
4	16	15	5

SOLUTION

We will compute all C_{ij} for H1 and will also show the computation procedure for C_{34} for H2 through H6. A Gantt chart representation of job 3 preceding job 4 is shown in Fig. 7-5. The following operation times are given in the example:

$$t(3, 2) = 13 \qquad t(3, 3) = 17$$
$$t(4, 1) = 16 \qquad t(4, 2) = 15$$

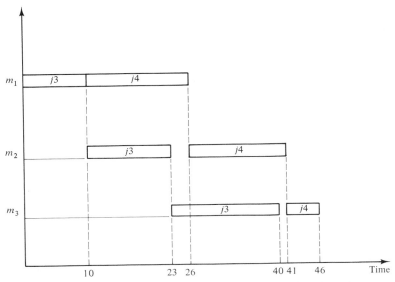

Figure 7-5. Gantt chart for Example 7-9 with $i = 3$ and $j = 4$.

From Eq. (7.11) we compute

$$r(34, 2) = 13 - 16 = -3$$
$$r(34, 3) = 17 - 15 = 2$$

From Eq. (7.14) we compute the following $r^*(ij, k)$ values:

$$r^*(34, 2) = 13 - 16 + 0 = -3$$
$$r^*(34, 3) = 17 - 15 - 3 = -1$$

For H1, using Eq. (7.12),

$$C_{34} = |-3| + |2| = 5$$

For H2, using Eq. (7.13),

$$C_{34} = (-3)^2 + (2)^2 = 13$$

For H3, using Eq. (7.15),

$$C_{34} = |-3| + |-1| = 4$$

For H4, using Eq. (7.16),

$$C_{34} = |-3| + |-1| = 4$$

For H5, using Eq. (7.17),

$$C_{34} = (-3)^2 + (-1)^2 = 10$$

For H6, using Eq. (7.18),

$$C_{34} = (0 + 2|-3|) + (2 + 2|0|) = 8$$

We now compute all C_{ij}'s for H1.

C_{12}:

$$r(12, 2) = 6 - 8 = -2$$
$$r(12, 3) = 15 - 11 = 4$$
$$C_{12} = |-2| + |4| = 6$$

C_{13}:

$$r(13, 2) = t(1, 2) - t(3, 1) = 6 - 10 = -4$$
$$r(13, 3) = t(1, 3) - t(3, 2) = 15 - 13 = 2$$
$$C_{13} = |-4| + |2| = 6$$

C_{14}:

$$r(14, 2) = t(1, 2) - t(4, 1) = 6 - 16 = -10$$
$$r(14, 3) = t(1, 3) - t(4, 2) = 15 - 15 = 0$$
$$C_{14} = |-10| + |0| = 10$$

C_{21}:

$$r(21, 2) = t(2, 2) - t(1, 1) = 11 - 14 = -3$$
$$r(21, 3) = t(2, 3) - t(1, 2) = 4 - 6 = -2$$
$$C_{21} = |-3| + |-2| = 5$$

C_{23}:

$$r(23, 2) = t(2, 2) - t(3, 1) = 11 - 10 = 1$$
$$r(23, 3) = t(2, 3) - t(3, 2) = 4 - 13 = -9$$
$$C_{23} = |1| + |-9| = 10$$

C_{24}:

$$r(24, 2) = t(2, 2) - t(4, 1) = 11 - 16 = -5$$
$$r(24, 3) = t(2, 3) - t(4, 2) = 4 - 15 = -11$$
$$C_{24} = |-5| + |-11| = 16$$

C_{31}:

$$r(31, 2) = t(3, 2) - t(1, 1) = 13 - 14 = -1$$
$$r(31, 3) = t(3, 3) - t(1, 2) = 17 - 6 = 11$$
$$C_{31} = |-1| + |11| = 12$$

C_{32}:

$$r(32, 2) = t(3, 2) - t(2, 1) = 13 - 8 = 5$$
$$r(32, 2) = t(3, 3) - t(2, 2) = 17 - 11 = 6$$
$$C_{32} = |5| + |6| = 11$$

C_{34}:

$$r(34, 2) = t(3, 2) - t(4, 1) = 13 - 16 = -3$$
$$r(34, 3) = t(3, 3) - t(4, 2) = 17 - 15 = 2$$
$$C_{34} = |-3| + |2| = 5$$

C_{41}:

$$r(41, 2) = t(4, 2) - t(1, 1) = 15 - 14 = 1$$
$$r(41, 3) = t(4, 3) - t(1, 2) = 5 - 6 = -1$$
$$C_{41} = |1| + |-1| = 2$$

Table 7-1. VALUES OF C_{ij} FOR H1

Predecessor	Follower, j			
i	1	2	3	4
1	∞	6	6	10
2	5	∞	10	16
3	12	11	∞	5
4	2	13	13	∞

C_{42}:

$$r(42, 2) = t(4, 2) - t(2, 1) = 15 - 8 = \quad 7$$

$$r(42, 3) = t(4, 3) - t(2, 2) = \quad 5 - 11 = -6$$

$$C_{42} = |7| + |-6| = 13$$

C_{43}:

$$r(43, 2) = t(4, 2) - t(3, 1) = 15 - 10 = \quad 5$$

$$r(43, 3) = t(4, 3) - t(3, 2) = \quad 5 - 13 = -8$$

$$C_{43} = |5| + |-8| = 13$$

These C_{ij} values are then placed into the matrix shown in Table 7-1. The main diagonal contains ∞, indicating an infeasible schedule option (i.e., no job may follow itself). The objective is to find the sequence of jobs that minimizes the total cost. Using the traveling salesman algorithm to solve the problem, we find that the optimal sequence is $4 \rightarrow 1 \rightarrow 2 \rightarrow 3$, which is identical to the one obtained by the branch-and-bound technique in Example 7-6.

7.8. Minimization of Setup Costs (Traveling Salesman Problem)

In all the sequencing problems presented so far, we have not considered separately the setup time for the machine and processing time of the job on the machine; that is, the machine time and setup time were totaled for every job to be presented by a single value t_{ij} (time to complete job i on machine j). However, there are many practical situations where the setup time for a job on a machine is dependent on the preceding job on that machine. For example, consider a paint making company that uses the same machine for mixing different paint colors. The preparation of the machine for each paint color depends on the color that was processed previously. Since the sum of processing times for all jobs is constant, while the setup times are dependent on the job sequence, an optimal job sequence is the one that minimizes the sum of setup times.

This problem is similar to the well-known traveling salesman problem in which a salesman, starting in one city, wishes to visit each $n - 1$ other cities once and only once and return to the starting city. In doing so, the salesman is seeking the optimal tour (the order of visiting the cities) that minimizes the total distance traveled. The distance between cities in the traveling salesman problem corresponds to the setup times of jobs on the machine in the sequencing problem. Little et al. [24] have presented a branch-and-bound approach for solving the traveling salesman problem. We utilize their approach to obtain the optimum sequence of jobs with dependent setup times as follows:

1. Construct a matrix $D = [d(i, j)]$, where the entry in row i and column j of the matrix is the setup time when job i immediately precedes job j. If $d_{ij} = d_{ji}$, the problem is said to be *symmetric*. On the other hand, if $d_{ij} \neq d_{ji}$, the problem is then referred to as *asymmetric*.
2. Perform a *row reduction* process on the matrix D by subtracting the smallest element in a row from every element of that row.
3. Perform a *column reduction* process in the same manner as the *row reduction* process. Note that steps 2 and 3 are interchangeable.
4. We refer to the matrix D at this stage as a *reduced matrix, D_r*. Suppose that S is the sum of the reducing constants from steps 2 and 3. The total setup time under the matrix D before reduction is given by

$$\text{LOT} = (\text{LOT})_r + S \qquad (7.19)$$

where $(\text{LOT})_r$ is the length of time to set up all jobs in the optimal sequence under the reduced matrix D_r. Since the reduced matrix D_r contains only nonnegative elements, S constitutes a lower bound on the length of the sequence under the original matrix D.
5. The branching process is similar to the one that has been presented earlier for the *n*-job, three-machine problem. Therefore, we construct a similar tree, as shown in Fig. 7-6. The starting node in the tree is called the *all-sequence node*. A node on the tree containing the job pair (i, j)

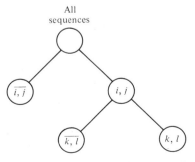

Figure 7-6. Start of a decision tree for the branch and bound algorithm.

represents all sequences, which include this pair (i, j). On the other hand, a node containing $\overline{(i, j)}$ represents all sequences that do not include (i, j).

6. The branching process continues at node (i, j) to nodes $\overline{(k, l)}$ and (k, l). The node $\overline{(k, l)}$ represents all sequences that include (i, j) but not (k, l), and node (k, l) represents all sequences which include both (i, j) and (k, l). We continue the branching process until a single sequence is obtained. The decisions as to which node to branch from and which pair of jobs should be included in the sequence are dependent on the lower bound on the nodes and the penalty for not including the pair of jobs in the sequence, respectively.

7. In this step we determine which pair of jobs are to be included in the sequence by evaluating the total setup time associated with the two possible alternatives at that branching stage. We define $LB(x)$ to be the lower bound on all sequences included in node x. The lower bound of the node *all sequences*, LB(all sequences), is the sum of the constants that were needed to reduce the original matrix D. Consider that we are branching to a new node Z. One possible job pair to be included in Z is the pair of jobs that have $d_{ij} = 0$ (minimum setup time). Now we consider all sequences that do not include the pair (i, j). Since job i must follow a job other than j and some other job will immediately follow i, any of these sequences must include a setup time at least equal to the sum of the smallest elements in row i and column j, excluding d_{ij}. Call the sum of these two setup times θ_{ij}.

8. After calculating θ_{ij} for each *zero* element in the reduced matrix, we choose to include in the sequence the job pair (i, j) with the largest penalty associated with not including it and the lower bound on the node $\overline{Z}$, which excludes (i, j), is

$$\text{LB}(\overline{i, j}) = \text{LB(all sequences)} + \theta_{ij} \tag{7.20}$$

Since the job pair (i, j) is included in the sequence, we branch to node Z and delete row i and column j from the matrix D_r. We also set $d_{ji} = \infty$ to avoid any subsequences. The resultant matrix is then reduced to form a lower bound for node Z as

$$\text{LB}(i, j) = \text{LB(all sequences)} + S_1 \tag{7.21}$$

where S_1 is the sum of reducing constants.

9. Repeat steps 7 and 8 until we arrive at a node representing a single sequence. If the lower bound on this node is less than or equal to the lower bounds on all other sequences, we have fathomed an entire branch and have an optimal solution. If not, we begin to fathom the nodes with minimum values on the lower bound for a possible optimal sequence.

These procedures are illustrated by Example 7-10.

EXAMPLE 7-10

Find the optimal solution to the following six-job, one-machine problem. The elements of the matrix represent the setup times.

				To Job j			
	1	2	3	4	5	6
1	∞	15	13	4	8	11
2	8	∞	12	8	10	6
From 3	5	7	∞	9	6	8
Job i 4	11	13	5	∞	8	7
5	11	5	8	10	∞	12
6	8	7	6	9	11	∞

SOLUTION

—Step 1: Perform a row reduction operation on the cost matrix.
—Step 2: Perform a column reduction operation on the cost matrix.
—Step 3: The reduced matrix is given in Table 7-2. Calculate S as the sum of the reducing constants from steps 1 and 2. $S = 32$.

Table 7-2. REDUCED MATRIX OF EXAMPLE 7-10

				To			
	1	2	3	4	5	6	
1	∞	11	9	0⑤	3	7	4
2	2	∞	6	2	3	0④	6
From 3	0②	2	∞	4	0②	3	5
4	6	8	0②	∞	2	2	5
5	6	0④	3	5	∞	7	5
6	2	1	0①	3	4	∞	6
			1				

—Step 4: Consider branching from the all-sequences node as shown in Fig. 7-7. [Note that LB(all sequences) = S = 32.]
—Step 5: Calculate θ_{ij} for each zero element of the reduced matrix, place it in the reduced matrix, and identify with circles. Since the max $\{\theta_{ij}\} = \theta_{14} = 5$, we branch from all sequences to node (1, 4). The lower bound on (1, 4) is found by Eq. (7.20) to be

$$\text{LB}(\overline{1, 4}) = 32 + 5 = 37$$

We delete row 1 and column 4 and set $d_{41} = \infty$ to avoid the formation of subsequences. By doing so, the reduced matrix becomes as shown in Table 7-3.
—Step 6: Repeat steps 1 through 5. The reduced matrix given in Table 7-3 cannot be reduced further. The lower bound on node (1, 4) is calculated using Eq. (7.21).

$$\text{LB}(1, 4) = 32 + 0 = 32$$

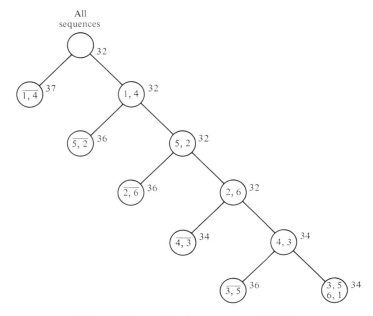

Figure 7-7. Decision tree for Example 7-10.

Since LB(1, 4) < LB($\overline{1, 4}$), we branch from node (1, 4). The θ_{ij}'s are the circled numbers in the matrix. Since max $\{\theta_{ij}\} = \theta_{26} = \theta_{52} = 4$, we have the option to branch from (1, 4) to either nodes (2, 6) and ($\overline{2, 6}$) or nodes (5, 2) and ($\overline{5, 2}$). We arbitrarily choose to branch from (1, 4) to (5, 2) and ($\overline{5, 2}$).

$$\text{LB}(\overline{5, 2}) = 32 + \theta_{52} = 32 + 4 = 36$$

We then delete row 5 and column 2 from the matrix and set $d_{25} = \infty$. The resultant matrix is shown in Table 7-4. Repeat steps 1 through 5. The new reduced matrix given in Table 7-4 cannot be reduced further; therefore,

$$\text{LB}(5, 2) = 32 + 0 = 32$$

Table 7-3. THE REDUCED MATRIX AFTER
DELETING ROW 1 AND COLUMN 4

		To				
		1	*2*	*3*	*5*	*6*
	2	2	∞	6	3	0④
	3	0②	2	∞	0②	3
From	*4*	∞	8	0②	2	2
	5	6	0④	3	∞	7
	6	2	1	0①	4	∞

Table 7-4. THE REDUCED MATRIX
AFTER DELETING ROW
5 AND COLUMN 2

		To			
		1	*3*	*5*	*6*
	2	2	6	∞	0④
From	3	0②	∞	0②	3
	4	∞	0②	2	2
	6	2	0②	4	∞

The new values of θ_{ij} are calculated and circled as shown in Table 7-4. Since θ_{26} is the largest θ_{ij}, we branch from node (5, 2) to $(\overline{2, 6})$ and (2, 6).

$$LB(\overline{2, 6}) = 32 + 4 = 36$$

Again, deleting row 2 and column 6 from the cost matrix and setting $d_{62} = \infty$, we get the matrix shown in Table 7-5. The matrix above cannot be reduced further. Therefore,

$$LB(2, 6) = 32 + 0 = 32$$

Calculate θ_{ij} and find max $\{\theta_{ij}\} = \theta_{35} = \theta_{43} = \theta_{31} = \theta_{63} = 2$. We choose to branch to $(\overline{4, 3})$ and (4, 3).

$$LB(\overline{4, 3}) = 32 + 2 = 34$$

Delete row 4 and column 3 in the matrix given in Table 7-5 to get the matrix shown in Table 7-6. Row 6 in the cost matrix can be reduced by 2 as shown in Table 7-6.

$$LB(4, 3) = 32 + 2 = 34$$

Finally, we branch from (4, 3) to (3, 5) and $(\overline{3, 5})$.

$$LB(\overline{3, 5}) = 34 + 2 = 36$$

$$LB(3, 5) = LB(6, 1) = 34$$

Table 7-5. THE REDUCED MATRIX
AFTER DELETING ROW
2 AND COLUMN 6

		To		
		1	*3*	*5*
	3	0②	∞	0②
From	4	∞	0②	2
	6	2	0②	4

Table 7-6. THE REDUCED MATRIX
AFTER DELETING ROW
4 AND COLUMN 3

$$To$$

		1	*5*	
From	*3*	$0^{\textcircled{0}}$	$0^{\textcircled{2}}$	
	6	$0^{\textcircled{2}}$	2	2

and this concludes the sequence. Since the lower bound on the last node is the smallest in the tree with the exception of (4, 3). Then the problem is solved and the optimum sequence is $1 \to 4 \to 3 \to 5 \to 2 \to 6$. The sum of the setup times of the optimum sequence is 34.

7.9. Job Shop Scheduling

We indicated earlier that the problems of production sequencing are quite complex due to the combinational difficulties. Although some progress has been made in the flow shop sequencing problem, little or insignificant progress has been made in the job shop sequencing problem. This is due to the large number of complexity factors that exist in the job shop environment, such as the fact that jobs do not have the same technological ordering. Because of the complexity of these problems, nearly all the reported studies have resorted to simulation as the research tool. In this section we present the most common scheduling (or loading) rules for the job shop environment. We also discuss dynamic job shop scheduling. The most commonly evaluated methods to schedule jobs (or operations) in a job shop production system are:

1. *RANDOM*:   Select the job at random.
2. *FCFS*:   Select the job on a first-come-first-served basis.
3. *DDATE*:   Select the job with the earliest due date.
4. *SPT*:   Select the job with the shortest processing time.
5. *LRPT*:   Select the job with the least remaining processing time.
6. *S/OPR*:   Select the job with the minimum ratio of job slack time to the number of operations remaining.
7. *ODD*:   Select the job with the earliest operational due date.
8. *LSTART*:   Select the job with the earliest late start time.
9. *OSLK*:   Select the job with the least operational slack.

These rules are self-explanatory. A complete definition of all the loading rules given above can be found in Farn and Muhlemann [11]. In general, some of these rules are considered to be superior to others. Clearly, this depends on countless factors, such as the measure of performance (criterion for comparison), level of uncertainty, and job arrival pattern.

There is a dynamic aspect of the job shop scheduling problem. A schedule for the current set of jobs is produced (by some method—discussed earlier in this chapter), and as this schedule is being worked through, new jobs arrive at the shop. Muhlemann et al. [26] states that there are two extremes for scheduling the new jobs: to produce a new schedule each time a new job arrives, or to completely finish the existing schedule before producing a new schedule for the jobs that had arrived in the meantime. The latter extreme changes the problem from dynamic job shop to static job shop. Alternatively, a solution some way between these two extremes is a possibility. Ignoring computational consideration, the *best* approach may be to have an on-line computer system whereby every time an operation finishes on a machine, this information is fed into the computer. The program then considers the actual status of the shop at that time and all relevant data and applies some criterion to select the next operation for that machine. Clearly, the computational implications of this are considerable and in many cases this is not economically feasible. Consequently, less frequent rescheduling must be considered. It would seem reasonable to expect overall performance to improve as rescheduling becomes more frequent. However, the precise extent of this and the computational cost are not fully understood and are of great interest to researchers.

7.10. Assembly Line Balancing

The *assembly line* is a production line where material moves continuously at a uniform average rate through a sequence of workstations where assembly work is performed. Typical examples of these assembly lines are car assembly, electric washers and dryers, electronic appliances, computer assemblies, and toy manufacturing and assembly. The assembly lines will continue to be a major part of many manufacturing and assembly operations, although its labor content may be reduced through robotization. A diagrammatic sketch of a typical assembly line is shown in Fig. 7-8. The arrangement of work along

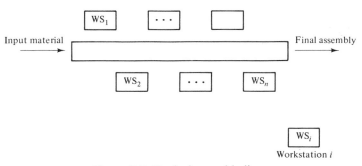

Figure 7-8. Typical assembly line.

the assembly line will vary according to the size of the product being assembled, the precedence requirements, the available space, the work elements, and the nature of the work to be performed on the job.

There are two main problems in assembly lines:

1. Balancing the workstations
2. Keeping the assembly line in continuous production

It is reported that downtime of an assembly line costs a major car manufacturer $98,000 per minute. *Perfect balance* of the line means to combine the elements of work to be done in such a manner that at each station the sum of the elemental times just equals the *cycle time*. Cycle time is the amount of time a unit of product being assembled is normally available to an operator performing the assigned task. Conveyors are the key material movers in most of the assembly lines. There are several types of conveyors that are used in assembly lines; the most widely used are belt, chain, overhead, pneumatic, and screw conveyors.

7.10.1. DEFINITIONS

The following definitions are applicable to all assembly lines.

1. *Assembled product:* The product that passes through a sequence of workstations where tasks are performed on the product until it is completed at the final workstation. The throughput of the assembly line is measured by the number of assembled products per unit time. The term "assembled product" should not imply that the completion of the product at the final station results in the end product (i.e., the assembled product could be a subassembly of a main product).
2. *Work element:* a part of the total work content in an assembly process. We define N as the total number of work elements required to complete the assembly and i is the work element number i in the process. Note that $1 \le i \le N$.
3. *Workstation (WS):* a location on the assembly line where a work element or elements are performed on the product. The minimum number of workstations, K, is greater than or equal to 1.
4. *Cycle time (CT):* The time between the completion of two successive assemblies, assumed constant for all assemblies for a given conveyor speed. The minimum value of the cycle time must be greater than or equal to the longest station time.
5. *Station time (ST):* It is the sum of the times of work elements which are performed at the same workstation. It is obvious that the station time (ST) should not exceed the cycle time (CT).
6. *Delay time of a station:* It is the difference between the cycle time (CT) and the station time (ST) (i.e., the idle time of the station = $CT - ST$).

7. *Precedence diagram:* A diagram that describes the ordering in which work elements should be performed (see Chapter 6). Some jobs cannot be performed unless its predecessors are completed. In fact, the layout of workstations along the assembly line depends on the precedence diagram.

The assembly line balancing problem can be stated as follows. Given a product that requires N elements of work to be completed, find the perfect balance of the assembly line. When the perfect balance cannot be achieved, we measure the effectiveness of the balance by:

1. *Line efficiency (LE):* it is the ratio of total station time to the cycle time multiplied by the number of workstations. It is expressed as

$$LE = \frac{\sum_{i=1}^{K} ST_i}{(K)(CT)} \times 100\% \tag{7.22}$$

where ST = station time of station i
K = total number of workstations
CT = cycle time

2. *Smoothness index (SI):* It is an index to indicate the relative smoothness of a given assembly line balance. A smoothness index of 0 indicates a perfect balance.

$$SI = \sqrt{\sum_{i=1}^{K} (ST_{max} - ST_i)^2}$$

where ST_{max} = maximum station time
ST_i = station time of station i
K = total number of workstations

In designing an assembly line, the following restrictions must be imposed on grouping of work elements:

1. Precedence relationship.
2. The number of workstations cannot be greater than the number of work elements (operations). Also, the minimum number of work stations is 1 (i.e., $1 \leq K \leq N$).
3. The cycle time is greater than or equal to the maximum time of any station time and of the time of any work element T_i. The station time should not exceed the cycle time, that is,

$$T_i \leq ST_i \leq CT$$

One method of seeking an optimal solution to the assembly line balancing problem is to start with the first station and select the combination of elements that will result in the least amount of idle time at that station (adhering to the restrictions imposed earlier) and then proceed to the next station, having eliminated from further consideration those elements already selected and to repeat this process until all elements are assigned. This process will not always lead to an optimum solution.

To date, there is no methodology that guarantees an optimal solution for all assembly line balancing problems. Researchers tend to use heuristic methods that can obtain a fairly good balance for the problems. Presented are three different methods for obtaining a good balance for the assembly line balancing problem.

7.10.2. KILBRIDGE–WESTER HEURISTIC METHOD

In the procedure proposed by Kilbridge and Wester [21], numbers are assigned to each operation describing how many predecessors it has. Operations with the lowest predecessor number are assigned first to the workstations. The following are the procedures to be followed as proposed by Kilbridge and Wester:

1. Construct the precedence diagram for the work elements. In the precedence diagram, list in column I all work elements that need not follow others. In column II, list work elements that must follow those in column I. Continue to the other columns in the same way. There exist many orderings that satisfy the precedence diagram. Movements between columns are totally free as long as the elements are not connected by arrows.

2. Cycle time CT is then determined by finding all combinations of the primes of $\sum_{i=1}^{N} T_i$ (the total elemental times). Select a feasible cycle time CT. The permissible number of stations is

$$ K = \frac{\sum_{i=1}^{N} T_i}{\text{CT}} $$

3. Assign work elements to the station such that the sum of the elemental times does not exceed CT.

4. Delete the assigned elements from the total number of work elements and repeat step 3.

5. If the station time exceeds CT due to the inclusion of a certain work element, this work element should be assigned to the next station.

6. Repeat steps 3 to 5 until all elements are assigned to workstations.

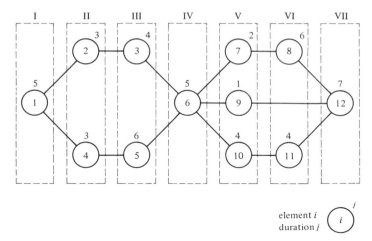

Figure 7-9. Precedence diagram for Example 7-11.

EXAMPLE 7-11

Use the Kilbridge–Wester method to balance the assembly line that has the precedence diagram shown in Fig. 7-9.

SOLUTION

We set

element 1	in column I
elements 2 and 4	in column II
elements 3 and 5	in column III
element 6	in column IV
elements 7, 9, and 10	in column V

Table 7-7. WORKSTATION ASSIGNMENTS

Column	Element i	T_i	Column Sum	Cumulative Sum
I	1	5	5	5
II	2	3		
	4	3	6	11
III	3	4		
	5	6	10	21
IV	6	5	5	26
V	7	2		
	9	1		
	10	4	7	33
VI	8	6		
	11	4	10	43
VII	12	7	7	50

elements 8 and 11   in column VI

element 12   in column VII

for this example.

$$\sum_{i=1}^{N} T_i = 50$$

Work Element (i)	T_i
1	5
2	3
3	4
4	3
5	6
7	2
8	6
9	1
10	4
11	4
12	$7(t_{max})$

$$\sum_{i=1}^{N} T_i = 50$$

The primes of 50 are $2 \times 5 \times 5$. The constraints on the cycle time are given by $7 \leq CT \leq 50$. Then all possible combinations of the primes are classified as yielding feasible or infeasible cycle times.

Feasible cycle times	Infeasible cycle times
$C_1 = 50$	$C_4 = 2$
$C_2 = 5 \times 5 = 25$	$C_5 = 5$
$C_3 = 2 \times 5 = 10$	

Arbitrarily, let us balance the line for $CT = 10$. The first feasible station design assignments are shown in Table 7-7. We count the number of predecessors for each work element as given in Table 7-8. We regroup the work elements in the workstations. Operation 1 is selected first, because it has the least number of predecessors. Therefore, we assign element 1 to station 1. Either element 2 or 4, each of which has an operation time of 3, can be assigned to station 1. We choose to assign element 2 to station 1, which results in a station time equals $8 \leq CT$. Element 4 cannot be added to station 1; otherwise, the station time will exceed the cycle time; therefore, we assign element 4 to station 2 and follow the process outlined above for assigning work elements to stations. Results of the assignment are shown in Table 7-9.

$$\text{line efficiency (LE)} = \frac{50}{6 \times 10} \times 100\% = 83.3\%$$

$$\text{smoothness index (SI)} = \sqrt{4 + 1 + 1 + 9 + 9} = \sqrt{24} = 4.89$$

Table 7-8. NUMBER
OF PREDECESSORS
FOR EACH WORK ELEMENT

Work Element i	Number of Predecessors	T_i
1	0	5
2	1	3
3	2	4
4	1	3
5	2	6
6	5	5
7	6	2
8	7	6
9	6	1
10	6	4
11	7	4
12	11	7

By examining Table 7-9 carefully, we can transfer elements between stations to obtain a better balance among workstations. Possible transfers are given in Table 7-10. Since the maximum station time $(S_{max}) = 9$, we can set the cycle time (CT) = 9. The LE and SI are

$$LE = \frac{50}{6 \times 9} \times 100 = 92.6\%$$

$$SI = \sqrt{1 + 1 + 1 + 1} = 2$$

The assignment given in Table 7-10 results in a significant improvement in the line efficiency and idle times of workstations.

Table 7-9. ASSIGNMENTS OF WORK ELEMENTS
TO STATIONS (CT = 10)

Station	Element i	T_i	Station Sum	$CT - ST_K$
I	1	5		
	2	3	8	2
II	4	3		
	5	6	9	1
III	3	4		
	6	5	9	1
IV	7	2		
	9	1		
	10	4	7	3
V	8	6		
	11	4	10	0
VI	12	7	7	3

Table 7-10. REVISED ASSIGNMENTS OF WORK ELEMENTS
TO STATIONS (CT = 9)

Station	Element i	T_i	Station Sum	$CT - ST_K$
I	1	5		
	2	3	8	1
II	4	3		
	5	6	9	0
III	3	4		
	6	5	9	0
IV	7	2		
	8	6	8	1
V	10	4		
	11	4	8	1
VI	9	1		
	12	7	8	1

7.10.3. MOODIE–YOUNG METHOD

This method consists of two phases.

—*Phase 1:* Work elements are assigned to consecutive workstations on the assembly line by the largest candidate rule. The largest candidate rule consists of assigning the available elements (those with no precedence restrictions) in order of declining time value. Hence, if two elements were available for assignment to stations, the one with the larger time value would be assigned first. After each element is assigned, available elements are considered in order of decreasing time value for the next assignment. We should use the matrix P (it indicates the immediate predecessors of each element) and matrix F (it indicates the immediate followers for each element) for the assignment procedure.

—*Phase 2:* Phase 2 attempts to distribute the idle time equally to all stations through the mechanism of trades and transfers of elements (adhering to the precedence constraint) between stations. The following are the steps of phase 2.

1. Determine both the largest and smallest station times from the balance of phase 1.
2. Call one-half the difference between these two values GOAL.

$$GOAL = \frac{ST_{max} - ST_{min}}{2}$$

3. Determine all single elements in ST_{max} which are less than twice the value of GOAL and will not violate precedence restrictions if transferred to ST_{min}.
4. Determine all possible trades of single elements from ST_{max} for single elements from ST_{min} such that the reduction in ST_{max} and subsequent gain in ST_{min} will be less than $2 \times$ GOAL.

5. Carry out the trade or transfer indicated by the candidate with the smallest absolute difference between itself and GOAL.
6. If no trade or transfer was possible between the largest and smallest stations, attempt trades and transfers between the ranked stations in the following order: with N (Nth-ranked station has greatest amount of idle time), $N - 1, \ldots, 3, 2, 1$.
7. If a trade or transfer is still not possible, drop the restrictions imposed by the value of GOAL and attempt, via the first six steps, to get a trade or transfer that will not increase the value of any station beyond that of the original cycle time.

EXAMPLE 7-12

Consider the 12-element problem given in Example 7-11. Balance the line using the Moodie–Young method.

SOLUTION

From Fig. 7-9 we first construct the P and F matrices as shown in Table 7-11. Column 1 of the P matrix is the listing of all element numbers. Columns 2, 3, and 4 of the matrix contain the work elements that immediately preceed the work element in column 1. Columns 2, 3, and 4 of the F matrix contain the elements that immediately follow the elements given in column 1. Perform the following steps:

1. Note the row of the P matrix which contains all zeros and assign the largest of the elements indicated by these rows if more than one exists.
2. Note the element numbers in the row of the F matrix which corresponds to the assigned element in step 1 and go to the rows of the P matrix that are indicated by these numbers; replace the assigned element's identification number with a zero.
3. Continue assigning elements following the procedure of steps 1 and 2 above and adhering to the restriction that

$$\max T_i \leq ST_K \leq CT$$

Table 7-11. P AND F MATRICES
FOR EXAMPLE 7-12

	P matrix				*F matrix*		
1	0	0	0	1	2	4	0
2	1	0	0	2	3	0	0
3	2	0	0	3	6	0	0
4	1	0	0	4	5	0	0
5	4	0	0	5	6	0	0
6	3	5	0	6	7	9	10
7	6	0	0	7	8	0	0
8	7	0	0	8	12	0	0
9	6	0	0	9	12	0	0
10	6	0	0	10	11	0	0
11	10	0	0	11	12	0	0
12	8	9	11	12	0	0	0

The problem is solved when the *P* matrix contains all zeros.

We now set CT = 10, and follow the steps above to obtain the assignment in Table 7-12.

$$LE = \frac{50}{6 \times 10} \times 100 = 83.3\%$$

$$SI = \sqrt{4 + 1 + 1 + 9 + 9} = 4.89$$

Now move to phase 2 to improve the solution of phase 1.

$$GOAL = \frac{10 - 7}{2} = 1.5$$

Transfer E_7 to ST_V and E_9 to ST_{VI} (E_7 must precede E_8 in ST_V). This will result in a maximum station time of 9, which reduces CT to 9.

$$LE = \frac{50}{6 \times 9} \times 100 = 92.6\%$$

$$SI = \sqrt{4} = 2$$

7.10.4. HELGESON–BIRNIE METHOD OR POSITIONAL WEIGHT TECHNIQUE

The steps involved in this technique are:

1. Develop the precedence network in the normal manner.
2. Determine the positional weight (PW) for each work element (a positional weight of an operation corresponds to the time of the longest path from the beginning of the operation through the remainder of the network).
3. Rank the work elements based on the positional weight in step 2. The work element with the highest positional weight is ranked first.

Table 7-12. END OF PHASE 1 ASSIGNMENT

Station	Element i	T_i	ST_K	Slack Time
I	1	5		
	2	3	8	2
II	4	3		
	5	6	9	1
III	3	4		
	6	5	9	1
IV	10	4		
	11	4		
	7	2	10	0
V	8	6		
	9	1	7	3
VI	12	7	7	3

Table 7-13. PW FOR OPERATIONS
OF EXAMPLE 7-13

Operation	Positional Weight, PW
1	34
2	27
3	24
4	29
5	25
6	20
7	15
8	13
9	8
10	15
11	11
12	7

4. Proceed to assign work elements (operations) to the workstations where elements of the highest positional weight and rank are assigned first.
5. If at any workstation additional time remains after assignment of an operation, assign the next succeeding ranked operation to the workstation, as long as the operation does not violate the precedence relationships, and the station time does not exceed the cycle time.
6. Repeat steps 4 and 5 until all elements are assigned to the workstations.

EXAMPLE 7-13

Use the *positional weight method* to balance the assembly line given in Example 7-11. Assume that CT = 10.

SOLUTION

We compute the positional weight for each operation as shown in Table 7-13. For example, the positional weight for operation 6 equals (PW of operation 6)

Table 7-14. RANK OF OPERATIONS
IN A DECREASING PW

Rank	Operation	PW
1	1	34
2	4	29
3	2	27
4	5	25
5	3	24
6	6	20
7	7	15
8	10	15
9	8	13
10	11	11
11	9	8
12	12	7

Table 7-15. ASSIGNMENTS OF THE WORK ELEMENTS

Station	Element E_i	T_i	ST_K	$CT - ST_K$
I	1	5		
	4	3	8	2
II	2	3		
	5	6	9	1
III	3	4		
	6	5	9	1
IV	7	2		
	10	4	6	4
V	8	6		
	11	4	10	0
VI	9	1		
	12	7	8	2

$$= \max \{(5 + 2 + 6 + 7), (5 + 1 + 7), (5 + 4 + 4 + 7)\} = 20.$$ We then rank the operations in a decreasing order of their positional weight (Table 7-14). Following steps 4, 5, and 6, we obtain the assignments of work elements shown in Table 7-15.

$$LE = \frac{50}{6 \times 10} \times 100 = 83.3\%$$

$$SI = \sqrt{4 + 1 + 1 + 16 + 4} = \sqrt{26} = 5.09$$

Trading E_{10} from ST_{IV} for E_8 from ST_V will reduce the cycle time to 9, increase the line efficiency to 92.6%, and reduce SI to 2.

7.11. Probabilistic Assembly Line Balancing

In the previous assembly line balancing methods, it is assumed that the work element times are invariant. There are many practical situations where the times of work elements are random variables which may or may not follow a probability distribution.

We present two approaches for balancing assembly lines when:

1. Work element times for station i follows a normal distribution with a mean μ_i and a standard deviation σ_i.
2. The probability distributions of the work element times are unknown; we know only μ_i and σ_i of the work element times of station i (i.e., it is a *distribution-free* case).

7.11.1. METHOD 1: THE PROBABILITY DISTRIBUTION IS NORMAL

In this method we utilize phase 1 of the Moodie–Young method for deterministic assembly line balancing with some modifications:

1. Construct P and F matrices (which are presented earlier).
2. Work elements are assigned to consecutive workstations on the assembly line by the largest candidate rule. This entails the assignment of

the available work elements in order of declining time value. Before assigning an element to the workstation, we must check step 3.

3. We state the following theorem.

If the distributions of two independent random variables have the means μ_1 and μ_2 and the variances σ_1^2 and σ_2^2, the distribution of their sum (or difference) has the mean $\mu_1 + \mu_2$ (or $\mu_1 - \mu_2$) and the variance $\sigma_1^2 + \sigma_2^2$.

Hence, if two work elements were available for assignment to a station, we calculate the probability that the station time will not exceed the cycle time if the first element is assigned to that station and compare with the probability that the station time will not exceed the cycle time due to the assignment of the second element to that station. The calculation of the probability is based on the theorem above. Find

$$z = \frac{\text{ST} - \text{CT}}{\sigma_{\text{station}}}$$

corresponding to the assignment of the first element to the station. Then find $P(\text{ST} < \text{CT})$ using the tables of standard normal distribution. If $P(\text{ST} < \text{CT}) \leq$ predetermined probability, assign this element to the station; otherwise, calculate $P(\text{ST} < \text{CT})$ for the second element.

4. After each element is assigned, available elements are reconsidered for the next assignment in order of decreasing time value and the probability that $\text{ST} < \text{CT}$ is less than or equal to a predetermined probability.

5. The steps above are repeated until all work elements are assigned to the work stations.

7.11.2. METHOD 2: DISTRIBUTION FREE

This method is similar to method 1, with the exception that we use Chebyshev's inequalities for estimating the probability that a station time will exceed the cycle time [27]. The following theorem is utilized.

If a probability distribution has any mean μ and standard deviation σ, the probability of obtaining a value that deviates from the mean by at least K standard deviations is at most $1/K^2$ for asymmetric distributions and $1/2K^2$ for symmetric cases. Symbolically:

For asymmetric distributions,

$$P(|X - \mu| > K\sigma) \leq \frac{1}{K^2} \qquad (K > 0)$$

For symmetric distributions,

$$P(|X - \mu| > K\sigma) \leq \frac{1}{2K^2}$$

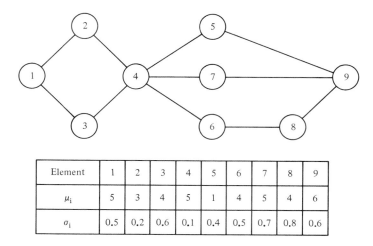

Element	1	2	3	4	5	6	7	8	9
μ_i	5	3	4	5	1	4	5	4	6
σ_i	0.5	0.2	0.6	0.1	0.4	0.5	0.7	0.8	0.6

Figure 7-10. Nine-work-element problem.

EXAMPLE 7-14

Under the following conditions, determine the allocation of work elements to the workstations for the nine work elements precedence diagram shown in Fig. 7-10.

1. The work elements are deterministic (i.e., $\sigma_i = 0$).
2. The work elements are probabilistic with mean μ_i and σ_i assuming that the predetermined probability ST > CT is set at two different values (0.0001 and 0.2).
   (a) The work elements follow a normal probability distribution.
   (b) The probability distribution of work elements is unknown.

SOLUTION

Allocations of elements to workstations are shown in Tables 7-16 through 7-20.

Table 7-16. NINE WORK ELEMENTS ASSEMBLY LINE
(deterministic)

Station	Element	T_K	ST_i	CT	Slack
1	1	5	5	13	8
	3	4	9		4
	2	3	12		1
2	4	5	5	13	8
	7	5	10		3
	5	1	11		2
3	6	4	4	13	9
	8	4	8		5
4	9	6	6	13	7

smoothness index = 8.88819   line efficiency = 0.711538

Table 7-17. NINE WORK ELEMENTS ASSEMBLY LINE
(normal distribution, $P = 0.0001$)

Station	Element	T_K	ST_i	CT	Slack
1	1	5	5	13	8
	3	4	9		4
2	2	3	3	13	10
	4	5	8		5
3	7	5	5	13	8
	6	4	9		4
4	8	4	4	13	9
	5	1	5		8
5	9	6	6	13	7

smoothness index = 13.0384   line efficiency = 0.569231

Table 7-18. NINE WORK ELEMENTS ASSEMBLY LINE
(normal distribution, $P = 0.2$)

Station	Element	T_K	ST_i	CT	Slack
1	1	5	5	13	8
	3	4	9	13	4
	2	3	12		1
2	4	5	5	13	8
	7	5	10		3
	5	1	11		2
3	6	4	4	13	9
	8	4	8		5
4	9	6	6	13	7

smoothness index = 8.88819   line efficiency = 0.711538

Table 7-19. NINE WORK ELEMENTS ASSEMBLY LINE
(distribution free, asymmetric
and $P = 0.20$)

Station	Element	T_K	ST_i	CT	Slack
1	1	5	5	13	8
	3	4	9		4
2	2	3	3	13	10
	4	5	8		5
	5	1	9		4
3	7	5	5	13	8
	6	4	9		4
4	8	4	4	13	9
	9	6	10	13	3

smoothness index = 7.54983   line efficiency = 0.711538

Table 7-20. NINE WORK ELEMENTS ASSEMBLY LINE
(distribution free, symmetric
and $P = 0.20$)

Station	Element	T_K	ST_i	CT	Slack
1	1	5	5	13	8
	3	4	9		4
2	2	3	3	13	10
	4	5	8		5
	5	1	9		4
3	7	5	5	13	8
	6	4	9		4
4	8	4	4	13	9
	9	6	10		3

smoothness index = 7.54983   line efficiency = 0.711538

7.12. Automatic Transfer Lines

The problem of designing automated production systems has received a great deal of attention in the literature. These production systems consist of a number of stages (arranged in series) at which operations are performed on the workpiece to produce a final product at the last stage. Operations at the stages are performed by machines or equipment, which are subject to failures (see references in [10]). Repairable failures of any stage will result in failure of the entire production system until repair is completed. Consequently, the production rate is decreased.

To improve the production rate of such systems, buffer storages of certain capacities are allocated between each pair of stages. The decision of how to allocate buffer storage to a production line is of practical importance to industry, especially those with assembly machines, and canning and packaging lines.

Allocation of buffer storage is not an easy task. In fact, it is impossible to find an allocation approach which determines the optimal size of buffers of a production system with a large number of production stages. The decision to estimate the size of a buffer can be governed by one or more of the following criteria:

1. What is the buffer size that maximizes the production rate of the system?
2. What is the buffer size that minimizes the total production cost?
3. What is the buffer size that maximizes the availability of the production system?

We now present methods for estimating the optimal buffer size.

Method 1: Consider a two-stage production system where the service rate (including downtime due to repair) is exponentially distributed with parameter μ units per unit time. The departure rate of units from the preceding stage is

Poisson distributed with a mean of λ units per unit time. The buffer size has a capacity of N units. One can view this system as a single-server queueing system with a finite queue length. (See any text on queueing theory or applications.) Suppose that our objective is to minimize the total production cost of the system. Let

C_s = storage cost per unit occupying the buffer per unit time

C_i = idle time cost per unit time

C_{sp} = cost of unit space per unit time

C_b = cost of blocking per unit time (blocking occurs when the first stage cannot deposit a finished unit because the buffer is full to its maximum capacity)

$TC(N)$ = total cost of production if N buffer spaces are allocated between stages of production

$E[N]$ = expected number of units waiting (being stored) in the buffer

P_b = probability that a stage is blocked

P_i = probability that a stage is idle

$\rho = \lambda/\mu$

$$TC(N) = C_i P_i + C_s E[N] + N C_{sp} + C_b P_b$$

where $E[N] = \dfrac{\rho}{1 - \rho} - \dfrac{(N + 1)\rho^{N + 1}}{1 - \rho^{N + 1}} + \dfrac{1 - \rho}{1 - \rho^{N + 1}} - 1$

$$P_i = \frac{1 - \rho}{1 - \rho^{N + 1}}$$

$$P_b = \frac{1 - \rho}{1 - \rho^{N + 1}} \rho^N$$

To find the N^* (optimal buffer size) that minimizes the total-cost equation, one should satisfy the following inequalities:

$$TC(N^* - 1) \geq TC(N^*) \leq TC(N^* + 1)$$

EXAMPLE 7-15

Determine the optimal buffer size of a two-stage transfer line. Production stages are independent

$$\lambda = 15 \text{ units/hr}$$

$$\mu = 20 \text{ units/hr}$$

$$C_s = \$0.10/\text{unit/hr}$$

$$C_{sp} = \$0.05/\text{hr}$$

$$C_b = \$15/\text{hr}$$

$$C_i = \$10/\text{hr}$$

SOLUTION

$$\rho = \frac{15}{20} = 0.75$$

$$P_i = \frac{0.25}{1 - (0.75)^{N+1}}$$

$$P_b = \frac{0.25}{1 - (0.75)^{N+1}} (0.75)^N$$

$$E[N] = \frac{0.75}{0.25} - \frac{(N + 1)(0.75)^{N+1}}{1 - (0.75)^{N+1}} + \frac{0.25}{1 - (0.75)^{N+1}} - 1$$

N	11	12	13
TC(N)	3.481	3.476	3.486

The optimal buffer size is $N^* = 12$.

Method 2: This method is based on the work of Panwalker and Smith [28]. There are K stages in series, each with identically independent exponential service time distribution of mean $(1/r)$. It is assumed that an infinite space in front of the first and last stages, which will result in no starvation of the first stage or blocking of the last stage. It is also assumed that there are n queue spaces (buffer spaces) between each pair of stages. The output rate at which units leave stage K after completing all services can be estimated as follows:

$$R_n(K) = r\left(A - \frac{C_K}{n + 3}\right)$$

where $R_n(K)$ = output rate of a system K stages and n buffer spaces between each pair of stages

$A \simeq 1$ for all values of n and K.

C_K = constant for a given K (the following table gives the value of this constant for selected K)

Number of Stages, K	C_K
3	1.356700
4	1.531307
5	1.631186
6	1.693860
7	1.736163
8	1.766183
9	1.787972
10	1.804640
15	1.848999
20	1.867132
25	1.874820
50	1.890346
100	1.890316

It should be noted that the value of C_K is estimated using a least-squares regression model.

EXAMPLE 7-16

What is the output rate of a three-stage series system with equal buffers of 10 units between stages? The service rate at each stage is 20 units per hour.

SOLUTION

$$R_{10}(3) = 20\left(1 - \frac{1.356700}{10 + 3}\right) = 17.9 \text{ units/hr}$$

PROBLEMS

7-1. (a) The following jobs are to be processed on a drill. Determine the sequence that minimizes the mean flow time.

Job	1	2	3	4	5	6	7
Processing Time	10	5	8	7	5	4	8

(b) Prove that the SPT rule minimizes the mean flow time of the jobs.

(c) Use the LPT (longest processing time) rule for sequencing jobs given in part (a). Compare the mean flow times of parts (a) and (c).

7-2. Assume that priorities are assigned to the jobs given in Prob. 7-1. The priority values are shown below.

Job	1	2	3	4	5	6	7
Priority	8	3	5	7	6	1	2

Find the best sequence such that the MFT is minimized (note that 8 is the highest value on the priority scale).

7-3. Prove that the sequence that minimizes the maximum lateness of N jobs to be processed on one machine is achieved by processing the jobs in order of nondecreasing due dates.

7-4. Eight jobs must be processed through a two-machine flow shop. The processing times of each job on both machines are shown below. Determine the makespan schedule.

Job	1	2	3	4	5	6	7	8
Processing time on M1	10	12	13	7	8	5	4	3
Processing time on M2	4	9	11	8	7	5	10	2

7-5. A set of 10 jobs to be processed on two machines, MA and MB. Each job must be processed on MA first, followed by MB. The processing times are as follows:

Job	1	2	3	4	5	6	7	8	9	10
MA	8	7	9	7	2	5	5	2	7	1
MB	6	5	2	8	4	1	7	3	3	5

Determine the minimum makespan schedule.

7-6. The following processing times were obtained for an N-job, three-machine problem. Find the minimum makespan schedule.

PROCESSING TIMES

Job	Machine 1	Machine 2	Machine 3
1	10	8	7
2	12	5	9
3	15	7	12
4	9	1	10
5	11	3	9
6	13	6	5

7-7. For technological reasons, jobs 4, 5, and 6 in Prob. 7-6 had to be replaced by job 5. The processing times of job 5 on M1, M2, and M3 are 10, 12, and 10, respectively. Find the new makespan schedule.

7-8. A three-machine job shop problem: Four jobs are to be processed through three machines. The following data are pertinent to the scheduling problem. Find the optimal sequence of the jobs.

PROCESSING TIMES

Job	Machine 1	Machine 2	Machine 3
1	7	10	8
2	5	7	9
3	11	5	9
4	8	9	8

7-9. Four jobs must be processed through five machines. The jobs pass through machines 1, 2, 3, 4, and 5 in the prescribed order. The technological ordering and processing times are shown below. Use the Campbell et al. algorithm to find the best schedule of the jobs. What is the minimum makespan obtained by this algorithm?

PROCESSING TIMES

Job	M1	M2	M3	M4	M5
1	5	7	8	6	9
2	10	12	5	4	9
3	11	11	7	3	8
4	21	9	19	4	1

7-10. Determine the lower bounds for the five jobs given in Prob. 7-9. Then use heuristic 1 (sum of absolute residuals) of the Stinson–Smith algorithm to determine the minimum makespan schedule. Compare the makespan with the lower bounds obtained earlier. Finally, compare the two makespans obtained in Probs. 7-9 and 7-10.

7-11. Three jobs to be processed on three machines, MA, MB, and MC, in the prescribed order. The processing times are:

PROCESSING TIMES

Job	MA	MB	MC
1	20	25	17
2	25	27	21
3	30	18	14

Find the minimum makespan schedule using
(a) The branch-and-bound algorithm.
(b) The Campbell et al. algorithm.
(c) The Stinson–Smith algorithm (H2).
(d) Compare the results of parts (b) and (c) with part (a).

7-12. Use the Stinson–Smith algorithm and apply criteria H1, H2, H3, H4, H5, and H6 to find the optimal schedule of the three jobs given in Prob. 7-11. Develop your own heuristic algorithm and compare results.

7-13. Two jobs are required to go through six machines. The processing times as well as the sequence of each job on the machines are:

see example
7-7 pg 245

Job 1 Sequence	A → B → D → C → F → E
Job 1 Times	5 6 4 7 3 4
Job 2 Sequence	B → A → D → C → E → F
Job 2 Times	7 5 6 8 2 1

Find the makespan of the optimal sequence.

7-14. The following are the setup and preparation times of a numerical control milling machine from job i to job j. What is the optimal schedule that minimizes the setup times for all jobs?

		To				
		1	2	3	4	5	6
	1	∞	10	17	12	15	10
	2	14	∞	15	6	7	8
	3	18	12	∞	21	15	11
From	4	11	16	13	∞	12	10
	5	20	17	11	6	∞	5
	6	10	6	7	5	10	∞

7-15. Develop your own heuristic algorithm to solve Prob. 7-14.

7-16. Assume that we have five jobs to be processed on one machine; the setup time of the machine for any job is dependent on the job that immediately precedes it. The setup times are given below. Find the optimal sequence.

		To			
		1	2	3	4	5
	1	∞	50	70	51	64
	2	71	∞	55	69	71
From	3	51	60	∞	82	85
	4	41	48	32	∞	75
	5	67	35	50	70	∞

7-17. *Scheduling for parallel processors:* Suppose that there are N jobs simultaneously available at time zero and M identical machines available for processing these jobs. A job processing can be interrupted and resumed on the same machine or a different machine without any loss in processing time. The objective is to determine the sequence of these jobs in order to minimize their makespan. The following heuristic rules could be utilized to determine such makespans.

1. *LRPT rule (longest remaining processing time):* Let

$P_j(t)$ = number of time periods required to process job j
r_j = job j will be ready to begin processing at the beginning of period r_j
d_j = due rate of job j
$R(t)$ = set of jobs at beginning of period t that are ready to be processed and still have some processing time remaining

$$R(t) = \{j/r_j \leq t \text{ and } P_j(t) > 0\}$$

The LRPT rule is summarized as follows:

At the beginning of period t for $t = 1, 2, \ldots, T$, if $R(t)$ contains M or fewer jobs, assign all jobs in $R(t)$ to the machines. If $R(t)$ contains more than M jobs, assign any M jobs to $R(t)$ such that if we assign job q and do not assign job r, then

$$P_q(t) \geq P_r(t)$$

2. *SRPT rule (shortest remaining processing time):* This rule is similar to LRPT except when $R(t)$ contains more than M jobs, we assign any M jobs in $R(t)$ such that if we assign job q and do not assign job r, then

$$P_q(t) \leq P_r(t)$$

Use the LRPT and SRPT rules to determine the schedule of the following jobs on two machines.

Job j	1	2	3	4	5	6
P_j	3	2	2	4	3	2
r_j	1	2	2	5	5	7
d_j	5	4	4	10	10	10

7-18. Use the DDATE rule (select the job with the earliest due date) to solve Prob. 7-17 and compare the makespans.

7-19. Find the number of workstations required to achieve balance for the following 10 work elements by using
(a) The Kilbridge–Wester method.
(b) The Moodie–Young method.

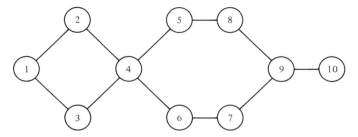

Figure 7-11. Precedence diagram for Prob. 7-19.

Assume a cycle time of 10 minutes.

Element i	1	2	3	4	5	6	7	8	9	10
Duration Time T_i	5	10	5	2	7	5	10	2	5	7

The precedence diagram is shown in Fig. 7-11.

7-20. Repeat Prob. 7-19 for a cycle time of 12 minutes.

7-21. Use the Helgeson–Birnie method to balance the assembly line whose precedence diagram is given in Prob. 7-19 (assume that CT = 10).

7-22. Under the following conditions, determine the allocation of work elements to the workstations for the 11 work elements precedence diagram shown in Fig. 7-12.
(a) The work elements are deterministic; that is, $\sigma_i = 0$.
(b) The work elements are probabilistics with mean μ_i and σ_i assuming that

   (1) The times of work elements follow a normal probability distribution and $P(ST > CT) = 0.01$.
   (2) The probability distribution of work elements is unknown.

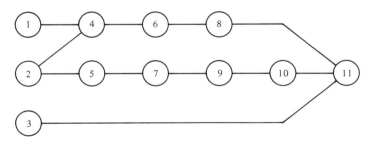

Element	1	2	3	4	5	6	7	8	9	10	11
μ_i	4	38	45	12	10	8	12	10	2	10	34
σ_i	0.1	0.5	0.7	0.3	0.4	0.3	0.6	0.8	0.1	0.3	0.5

Figure 7-12. Precedence diagram for Prob. 7-22.

ELEVEN WORK ELEMENTS PROBLEM

Element i	μ_i	σ_i
1	4	0.1
2	38	0.5
3	45	0.7
4	12	0.3
5	10	0.4
6	8	0.3
7	12	0.6
8	10	0.8
9	2	0.1
10	10	0.3
11	34	0.5

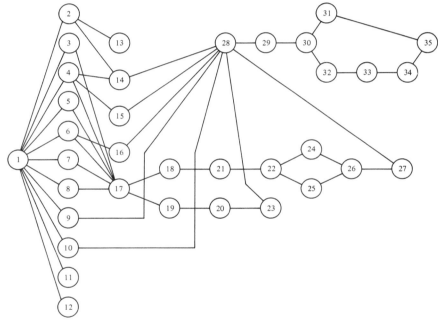

Element	1	2	3	4	5	6	7	8	9	10	11	12	13	14	15	16	17	18
μ_i	10	11	6	28	4	18	11	24	3	7	8	8	11	12	24	10	15	3
σ_i	0.1	0.3	0.5	0.9	1.1	0.2	0.6	0.4	0.8	1.0	0.15	0.7	0.55	0.9	0.3	0.2	0.8	0.2

Element	19	20	21	22	23	24	25	26	27	28	29	30	31	32	33	34	35
μ_i	19	63	10	11	13	8	8	33	11	21	6	8	17	14	11	9	16
σ_i	0.25	0.3	0.9	0.45	0.6	1.2	0.85	0.9	0.65	0.2	0.45	0.7	0.9	0.5	0.85	0.75	1.0

Figure 7-13. Precedence diagram for Prob. 7-23.

7-23. Repeat Prob. 7-22 for the 35-element problem given below. The precedence diagram is shown in Fig. 7-13.

Element	μ_i	σ_i	Element	μ_i	σ_i
1	10	0.1	19	19	0.25
2	11	0.3	20	63	0.3
3	6	0.5	21	10	0.9
4	28	0.9	22	11	0.45
5	4	1.1	23	13	0.6
6	18	0.2	24	8	1.2
7	11	0.6	25	8	0.85
8	24	0.4	26	33	0.9
9	3	0.8	27	11	0.65
10	7	1.0	28	21	0.2
11	8	0.15	29	6	0.45
12	8	0.7	30	8	0.7
13	11	0.55	31	17	0.9
14	12	0.9	32	14	0.5
15	24	0.3	33	11	0.85
16	10	0.2	34	9	0.75
17	15	0.8	35	16	1.0
18	3	0.2			

7-24. What is the implied cost of unit space per unit time of a buffer between stages of production in a typical transfer line? The following information is given:

$$\lambda = 18 \text{ units/hr}$$
$$\mu = 24 \text{ units/hr}$$
$$C_s = \$0.12/\text{unit/hr}$$
$$C_b = \$17/\text{hr}$$
$$C_i = \$12/\text{hr}$$

Optimal buffer size $N^* = 14$ units of space.

7-25. Determine the optimal buffer size of a two-stage production system. Assume that the stages are independent

$$\lambda = 10 \text{ units/hr}$$
$$\mu = 20 \text{ units/hr}$$
$$C_s = \$0.14/\text{unit/hr}$$
$$C_{sp} = \$0.10/\text{unit/hr}$$
$$C_b = \$16/\text{hr}$$
$$C_i = \$14/\text{hr}$$

7-26. What is the output rate of a 10-stage series system with equal buffers of 15 units between stages? Assume a service rate at each stage to be 40 units per hour.

REFERENCES

[1] AKERS, JR. SHELDON, B., AND JOYCE FRIEDMAN, "A Non-Numerical Approach to Production Scheduling Problems," *Operations Research*, 3, no. 4, 1955.

[2] BAKER, KENNETH, *Introduction to Sequencing and Scheduling*, New York: John Wiley and Sons, 1974.

[3] BAKER, KENNETH, AND WILLIAM BERTRAND, "A Comparison of Due Date Selection Rules," *AIIE Transactions*, Vol. 13, No. 2, June 1981.

[4] BECKMAN, R. R., "Sequencing of Two Jobs on M Machines," unpublished M.Sc., Arizona State University, Arizona, 1964.

[5] BLACKSTONE, JOHN H., JR., DON T. PHILLIPS, AND GARY L. HOGG, "A State of the Art Survey of Dispatching Rules for Manufacturing Job Shop Operations," *International Journal of Production Research*, Vol. 20, No. 1, Jan./Feb. 1982.

[6] BROOKS, GEORGE H. AND CHARLES R. WHITE, "An Algorithm for Finding Optimal or Near Optimal Solutions to the Production Scheduling Problem," *Journal of Industrial Engineering*, 16, no. 1, 1965.

[7] CAMPBELL, HERBERT G., RICHARD A. DUDEK, AND MILTON L. SMITH, "A Heuristic Algorithm for the n Job, M Machine Sequencing Problem," *Management Science*, 16, no. 10, (June 1970) pp. 630–637.

[8] CONWAY, RICHARD W., WILLIAM L. MAXWELL, AND LOUIS W. MILLER, *Theory of Scheduling*. Reading Mass.: Addison-Wesley Pub. Co., 1967.

[9] DUDEK, RICHARD A., MILTON L. SMITH, AND SHRIKANT S. PANWALKER, "Use of a Case Study in Sequencing/Scheduling Research," *OMEGA*, 2, 253, 1974.

[10] ELSAYED, ELSAYED A., AND RICHARD E. TURLEY, "Reliability Analysis of Production Systems With Buffer Storage," *International Journal of Production Research*, 5, no. 5 (1980) pp. 637–645.

[11] FARN, C. K., AND A. P. MUHLEMANN, "The Dynamic Aspects of a Production Scheduling Problem," *International Journal of Production Research*, 17, no. 15, 1979.

[12] GIGLIO, RICHARD J., AND HARVEY M. WAGNER, "Approximate Solutions to the Three Machine Scheduling Problem," *Operations Research*, 12, no. 2, 1964.

[13] GODIN, VICTOR B., "Interactive Scheduling: Historical Survey and State of the Art," *AIIE Transactions*, Vol. 10, No. 3, Sept. 1978.

[14] HARDGRAVE, W. W., AND G. L. NEMHAUSER, "Geometric Model and Graphical Algorithms for a Sequencing Problem," *Operations Research*, 11, no. 6, 1963.

[15] HELGESON, W. P., AND D. P. BIRNIE, "Assembly Line Balancing Using the Ranked Positional Weight Technique," *Journal of Industrial Engineering*, 6, no. 6, 1961.

[16] IGNALL, EDWARD J., "A Review of Assembly Line Balancing," *Journal of Industrial Engineering*, 16, no. 4, 1965.

[17] IGNALL, EDWARD J., AND LINUS E. SCHRAGE, "Application of the Branch and Bound Technique to Some Flow-Shop Scheduling Problems," *Operations Research*, 13, no. 3, 1965.

[18] JACKSON, JAMES R., "An Extension of Johnson's Results on Job-Lot Scheduling," *Naval Research Logistics Quarterly*, 3, no. 3, 1956.

[19] JOHNSON, S. M., "Optimal Two- and Three-Stage Production Schedules with Setup Times Included," *Naval Research Logistics Quarterly*, 1, no. 1, 1954.

[20] JOHNSON, LYNWOOD A., AND DOUGLAS C. MONTGOMERY, *Operations Research in Production Planning, Scheduling and Inventory Control*. New York, N.Y.: John Wiley & Sons, 1974.

[21] KILBRIDGE, MAURICE D., AND LEON WESTER, "A Heuristic Method of Assembly Line Balancing," *Journal of Industrial Engineering*, 12, no. 4, 1961.

[22] KING, J. R., AND A. S. SPACHIR, "Heuristics for Flow-Shop Scheduling," *International Journal of Production Research*, Vol. 18, No. 3, 1980.

[23] KOTTAS, JOHN F. AND HON-SHIANG LAU, "A Stochastic Line Balancing Procedure," *International Journal of Production Research*, Vol. 19, No. 2, Mar./Apr. 1981.

[24] LITTLE, JOHN D., KATTA G. MURTY, DURA W. SWEENEY, AND CAROLINE KAREL, "An Algorithm for the Traveling Salesman Problem," *Operations Research*, 11, no. 6, 1963.

[25] MOODIE, COLIN L., AND HEWITT H. YOUNG, "A Heuristic Method of Line Balancing for Assumptions of Constant or Variable Work Element Times," *The Journal of Industrial Engineering*, XVI, no. 1, Jan–Feb. 1965.

[26] MUHLEMANN, A. P., A. G. LOCKETT, AND C. K. FARN, "Job Shop Scheduling Heuristics and Frequency of Scheduling," *International Journal of Production Research*, 20, 1982.

[27] OMAR, MOHAMED T., "Development of a Heuristic Method for Assembly Line Balancing," unpublished M.Sc. Thesis, University of Windsor, 1975.

[28] PANWALKER, S. S. AND MILTON L. SMITH, "A Predictive Equation for Average Output of K Stage Series Systems with Finite Interstage Queues," *AIIE Transactions*, 11, no. 2, 1979.

[29] STARR, MARTIN, K., *Systems Management of Operations*. Englewood Cliffs, N.J.: Prentice-Hall, Inc., 1971.

[30] STINSON, JOEL P., AND ARTHUR W. SMITH, "A Heuristic Programming Procedure For Sequencing the Static Flowshop," *International Journal of Production Research*, 20, no. 6, 1982.

[31] SZWARC, WLODZIMIERZ, "The Flow-Shop Problem With Mean Completion Time Criterion," *IIE Transactions*, Vol. 15, No. 2, June 1983.

NEW DIRECTIONS IN BATCH
AND DISCRETE-PARTS PRODUCTION SYSTEMS

8.1. Introduction

Throughout the text we have emphasized the relevance of production control techniques to the specific system under consideration. This is especially true in Chapters 4 and 5, in which we contrast planning methods in the relatively simple product layout facility for mass production versus the relatively complex process layout plant for discrete-parts production. The control vehicle in the latter case is the MRP system with capacity and shop floor planning aids. MRP was developed so that production management could obtain a measure of control over the discrete parts production planning problem and also reduce work-in-process inventory. However, there are drawbacks to MRP in that it is very expensive to implement, requiring the capability of a mainframe computer, technical support professionals, and MRP software. This has led some manufacturers to take a contrary approach: instead of designing production control tools for a complex production system, they simplify the system. By so doing, they minimize the difficult problem of production control.

In this chapter we describe some of these production system design strategies and their implications for production control. In some cases, these new directions simplify or eliminate existing problems; in other cases, they eliminate some problems but create new and challenging ones.

8

8.2. Kanban

Kanban was developed in Japanese manufacturing, the best-known example being the Takahama plant of Toyota in Japan. Since then it has been adapted by some American manufactures. As practiced at Toyota, the *Kanban production system* is a noncomputerized production control system that works from a master schedule, just as an MRP system does. However, there are significant differences in the form that the master schedule takes as well as the way in which the production plan is executed.

An important difference occurs in production lot sizing. In computing production lot sizes, as described previously in this text, one tries to balance setup and carrying costs. When machine setup cost is high, which is typical in batch metalworking operations, this encourages relatively long production runs of a given product. Relatively long production runs implies long production lead times in the system and, consequently, more in-process inventory to be monitored and controlled over a longer period of time.

Kanban emphasizes the reduction in production lead time and in-process inventory by specifying shorter production runs of any single product. Figure 8-1 is an illustration of the master production schedule under conventional batch manufacture and Kanban, showing Kanban's emphasis on frequent setups and shorter production runs of a given product.

					Production Day						
	1	*2*	*3*	*4*	*5*	*6*	*7*	*8*	*9*	*10*	*Total*
					Conventional						
Product *Q* 204	100	100	100	100							400
Product *Z* 102					200	200	200	200			800
Product *Y* 101									150	150	300
					Kanban						
Product *Q* 204	100		100			100		100			400
Product *Z* 102		200		200			200		200		800
Product *Y* 101					150					150	300

Figure 8-1. Illustrated conventional and Kanban master schedules.

In batch manufacture, the time profile of inventory completing any stage of production looks like that of Fig. 8-2. For a batch of size Q, and assuming stockouts are negligible, the inventory cost is given by

$$K(Q) = \frac{AD}{Q} + h\frac{Q}{2} + h(\text{SS}) \tag{8.1}$$

where $K(Q)$ = total annual cost
$\quad\quad A$ = setup cost
$\quad\quad D$ = annual demand
$\quad\quad Q$ = batch size
$\quad\quad h$ = holding cost
$\quad (\text{SS})$ = average annual safety stock level

It is clear from Eq. (8.1) that the Kanban system, with small Q, produces lower cycle stocks of inventory. In addition, assuming that a continuous review system is used and demand is normally and independently distributed, the safety stock is given by

$$(\text{SS}) = Z_\alpha \sqrt{\sigma_D^2 L} \tag{8.2}$$

where Z_α = number of standard deviations associated with a level of protection α
$\quad \sigma_D^2$ = variance of demand
$\quad\quad L$ = lead time

Because the lead time L under Kanban is smaller than under conventional production systems, the variance of the lead time demand, $\sigma_D^2 L$, is smaller and the safety stock and its associated carrying cost is smaller.

Of course, if the conventional batch manufacturing system and the Kanban system have the same setup times, we would expect the total setup

Figure 8-2. Profile of batch manufactured inventory completing a stage of production.

cost, as given by (AD/Q), to be much higher under Kanban. However, a key and necessary ingredient in a Kanban system is low setup times and cost. Users of Kanban invest considerable time and effort into engineering *quick-change* tooling to reduce setup times. For example, at Toyota, a bolt maker that originally took 8 hours to change over was subsequently reduced to 58 seconds [13]. Hence a total reduction in inventory cost is achieved, driven by a very low setup cost and subsequent reductions in lot size and lead time. As we shall see next, it is impossible to control and economically run a Kanban system without extremely short lead times accomplished by quick changeover and small lot sizes.

8.2.1. INVENTORY CONTROL AND OPERATIONS SCHEDULING UNDER KANBAN

Inventory levels and operations scheduling in a Kanban system are driven by a set of cards, called *Kanban cards*. There are two kinds of Kanban cards used by Toyota, the *requisition card* and the *production card*. The requisition card authorizes withdrawal of material by an operation from a lower operation that feeds it. A production card authorizes the feeding operation to produce more of what is being requested.

The operation of the production system using Kanban cards is illustrated in Fig. 8-3. The master schedule dictates the schedule for final assembly. The assembly department knows the schedule for at least a week or two in advance and it also knows the lead times required to obtain lower-level components and subassemblies. Once again, these lead times must be extremely short. In order to have the required components for a particular day's production, final assembly will issue a Kanban requisition card one lead time prior to that day. In effect, this is the same kind of lead-time offsetting and scheduling performed within the centralized MRP system; in a Kanban system it is decentralized to the department level. Manufacturing maintains a buffer of component inventory which is intentionally kept very low. When component production is required, a Kanban production card is released to the relevant machining center. Manufacturing usually deals with the production of components on a

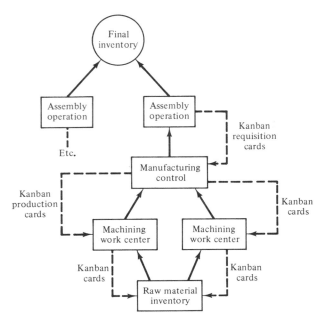

Figure 8-3. Illustrated Kanban system information and product flow. Material flow, solid line; information flow, dashed line.

first-come-first-served basis. When conflicts arise, management judgment is applied. Because lot sizes are small and setup times are short, machines are not tied up for a long time on the production of any one component; hence conflicts are minimized.

In summary, a Kanban production scheduling system is driven off a master schedule, as is MRP. However, within the system production control is decentralized. Production activity is regulated by Kanban cards, which are used to requisition lower-level requirements and initiate production. Conflicts are handled by management and supervisory intervention on the shop floor.

8.2.2. SOME ISSUES IN THE IMPLEMENTATION OF KANBAN

The simplicity of the Kanban system masks some important and difficult issues in its operation. Kanban requires a considerable amount of shop floor teamwork in decision making to ensure its success. Decentralized decision making in production is not characteristic of the mode of operation in many industrial firms. Consequently, in order to use Kanban, there may be considerable investment in retraining and redefining the roles of shop floor personnel.

Kanban does not work well in a highly engineered, one-at-a-time environment. Although production runs are short under Kanban, there must be some repetition in product manufacture.

Kanban does not provide methods for capacity planning. When a particular master schedule results in overloaded capacity, adjustments have to be made during production. This contrasts with a typical **MRP** planning system, where capacity requirements are projected through the planning horizon.

8.3. Group Technology

Group technology (GT) is a manufacturing philosophy in which similar parts are identified and grouped together to take advantage of their similarities in design and manufacture. Hence components requiring primarily external turning operations, such as shafts, are collected in one group, while components requiring surface grinding and drilling operations, such as plates, are assigned to a different group. These groups become the basis on which production engineers can reorganize a traditional process plant layout into a group technology plant layout, in which machines are arranged such that each machine is assigned to the production of only one group of parts.

Figure 8-4 illustrates the difference between the configuration of a process layout and GT. Through the simplification of routings, the group technology organization of production transforms a difficult job shop production control problem into a more simplified series of flow shop problems. The machine groupings, depending on how they are arranged, are usually referred to as either *group cells* or *group flow lines*. As opposed to Kanban, group technology typically affects only component manufacture, not the assembly stage of production.

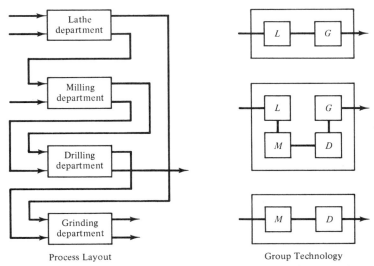

Process Layout   Group Technology

Figure 8-4. Product flow under process layout production organization and group technology.

One of the major benefits of the group production method lies in the dramatic reduction in setup times that are made possible through the use of group jigs. As a consequence of grouping components together based on similar machining requirements and component geometry, production engineers and tool designers are able to design fixtures specifically for ease of changeover among components within the group. In actual applications of GT, group jigs and associated changes have resulted in setup-time reductions in the order of 70% [8].

Another major benefit occurs in a reduction in component production lead time. In the process layout configuration, considerable time is spent transporting batches of product between machining departments. In addition, batches are queued at machining centers waiting to be released to an available machine. In the group flow line configuration, transportation is eliminated and queueing is minimal. In industrial applications of GT, companies have reported component production lead time reductions in the order of 70% [8].

The influence of setup-time and lead-time reductions on component inventories can be analyzed through an application of the standard inventory model to Fig. 8-2, which could represent the behavior of finished component cycle stock and safety stock. For finished component cycle stock, let

Q_p, Q_g = lot sizes under process layout and group technology organization, respectively

A_p, A_g = setup cost under process layout and group technology, respectively

R_{CS} = fractional reduction in average cycle stock due to a conversion from process to group technology layout; $R_{CS} = 1 - (Q_g/Q_p)$

D = annual demand rate

h = average inventory holding cost per unit

Then

$$Q_p = \sqrt{\frac{2A_p D}{h}}$$

$$Q_g = \sqrt{\frac{2A_g D}{h}}$$

$$R_{CS} = 1 - \sqrt{\frac{A_g}{A_p}} \tag{8.3}$$

assuming that the unit holding cost is equivalent under both configurations.[1] Figure 8-5 illustrates the relative component cycle stock reduction for various ratios of setup cost. If the use of group jigs reduces setup time (and cost) by 50%, finished component cycle stocks will be reduced by about 30%.

---

[1]For a more complete analysis, see Boucher and Muckstadt [3].

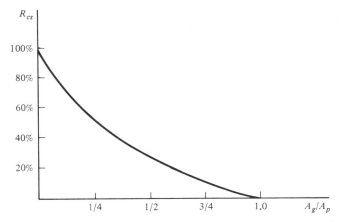

Figure 8-5. Reduction in cycle stock for various ratios of setup cost.

For the case of safety stocks, we define:

L_p, L_g = production lead times under process layout and group technology, respectively

r_p, r_g = reorder points under process layout and GT

Z_α = number of standard deviations associated with a level of protection α during the lead time

R_{SS} = fractional reduction in safety stock resulting from a conversion to group technology

σ_D^2 = variance of demand per year

Then assuming that a continuous review system is used and that demand is independently and normally distributed,

$$r_p = L_p D + Z_\alpha \sqrt{L_p \sigma_D^2}$$
$$r_g = L_g D + Z_\alpha \sqrt{L_g \sigma_D^2}$$

where $Z_\alpha \sqrt{L_p \sigma_D^2}$ and $Z_\alpha \sqrt{L_g \sigma_D^2}$ are the safety stocks of finished components for process and GT production organizations, respectively. Therefore,

$$R_{SS} = 1 - \sqrt{\frac{L_g}{L_p}} \tag{8.4}$$

Finally, component work-in-process inventory reductions are directly related to production lead-time reductions. We define:

M = raw material cost per unit for component manufacture

C = finished component manufacturing cost per unit

$v = C - M$ = unit value added in manufacture

Therefore, we can define the average dollar value of a batch of component work-in-process inventory as

$$\text{average WIP} = \left(M + \frac{v}{2} \right) Q$$

The average annual dollar value of work-in-process is the time average taken over the length of a production lead time times the number of cycles per year.

$$\text{average annual WIP} = \left[\int_0^L \left(M + \frac{C - M}{2} \right) Q \, dt \right] \frac{D}{Q}$$

$$W_A = \left(\frac{M + C}{2} \right) L D \tag{8.5}$$

where W_A is the dollar value of the annual average work-in-process inventory.

Assuming that $C_g = C_p$ and that $L_g < L_p$, the fractional reduction of work-in-process, R_{WIP}, is given by

$$R_{\text{WIP}} = 1 - \frac{L_g}{L_p} \tag{8.6}$$

For a 70% reduction in lead time, work-in-process inventory is reduced by 70%.

These simple computations illustrate the influence that changing the underlying production organization has on the inventory component of the production control problem. In addition, it should be noted that group technology usually yields other important benefits, such as:

—Simplification of complex planning problems through use of simplified flow line production organization
—Reduced percentage of defective parts
—Reduction in overall production floor space requirements
—Reduction of overdue orders as a result of much shorter production lead time

8.3.1. LOT SIZING UNDER GROUP TECHNOLOGY

We have just shown how analysis using the simple EOQ model gives a means of estimating the order of magnitude of the difference in inventory levels between process and group production organizations. In practice, this method of analysis would have to be tailored for the actual production system under consideration.

The single product EOQ model is often employed for sizing production lots in batch manufacturing; the model has also been extended to the case where several products share a common production facility. The EOQ model and its extensions optimize lot sizes by minimizing the sum of setup cost and finished-product carrying cost.

In group technology production there is a direct and traceable relationship between lot size and work-in-process inventory which can be incorporated into the lot sizing decision. Figure 8-6 illustrates the functional relationship between production lead time and lot size. Here we assume that queueing is negligible, although the results that follow are not sensitive to that assumption. From Eq. (8.5), we estimated W_A as

$$W_A = \left(M + \frac{v}{2} \right) LD$$

As shown in Fig. 8-6, $L = S_u + mQ$, where S_u is the total setup time for a batch and m is the total machining time per unit. Letting i represent the carrying cost rate, the annual carrying cost for work-in-process, $K(W)$, is

$$K(W) = i \left(M + \frac{v}{2} \right) (S_u + mQ)D$$

Adding setup cost and finished-goods carrying cost, we obtain the following total-cost function:

$$K(Q) = \frac{AD}{Q} + iC \frac{Q}{2} + i \left(M + \frac{v}{2} \right) (S_u + mQ)D$$

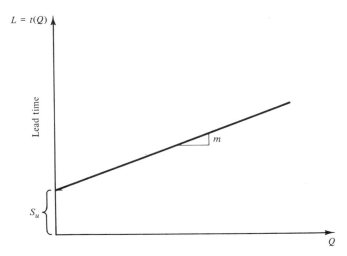

Figure 8-6. Total lead time L as a function of total setup time S_u and unit machining time m.

In a typical manufacturing situation, unit cost C and unit value added v are not constants. Since there is a fixed setup cost, both depend on the quantity produced. We define $C(Q)$ and $v(Q)$ as follows:

$$v(Q) = \left(\frac{S_u}{Q} + m\right)R$$

$$C(Q) = M + v(Q)$$

where R is the rate charged per unit of cell production time. R includes worker wages and cell overheads. R is in units of dollars per year. This results in a cost function of the form

$$K(Q) = \frac{AD}{Q} + i\frac{Q}{2}\left(M + \frac{S_u}{Q}R + mR\right) + iD\left(M + \frac{S_u}{2Q}R + \frac{m}{2}R\right)(S_u + mQ)$$

$$(8.7)$$

Differentiating with respect to Q yields the following minimum-cost lot size:

$$Q^* = \sqrt{\frac{2AD + DS_u(iS_uR)}{i(M + mR) + 2iDm(M + (mR)/2)}} \qquad (8.8)$$

Equation (8.8) illustrates the relationship to the standard EOQ formulation. The first terms under the radical in both the numerator and denominator are the usual expressions in the standard EOQ formula, with $M + mR$, the marginal unit production cost, replacing C. The second set of terms pertain to the influence of work-in-process inventory.

Boucher [2] has analyzed the effect on lot sizing of using Eq. (8.8), as opposed to the EOQ formula, in computing production lot sizes. In general, Eq. (8.8) yields significantly different lot sizes and lower cost primarily in cases where demand is moderately high and/or there is extensive component machining time.

8.3.2. OPERATIONS SCHEDULING UNDER GROUP TECHNOLOGY

Since a group technology production system can be configured as a flow line, there is the possibility of adopting the flow shop scheduling methods of Chapter 7 to GT. Indeed, this has been done by Hitomi and Ham [15], who apply mathematical programming and the branch-and-bound procedure to maximize the group component production rate for the single-machine case.

For larger, more complex group technology systems, efficient scheduling algorithms do not exist. The modeling problem in larger systems differs from that of the flow shop model in that the number of machines in the cell or line will probably exceed the number of workers; hence, at any point in time, some machines must be idle. Group technology cells or flow lines are purposely manned that way because there are usually machines with high levels of utilization and machines with very low levels of utilization within the same cell. In practice, it is often the case that the operating discipline within the cell is to have one worker assigned one job at a time which the worker takes sequentially through all its required machining operations. This operating discipline differs from that which is assumed by the flow shop models of Chapter 7.

In actual applications of GT, the following two scheduling disciplines are often found. The first is shop floor *supervisory scheduling*, where the cell supervisor chooses the next job to assign to a worker/machine combination based on machine availability and some priority scheme, such as earliest due date. This method is used especially where work rules prevent the training of machine operators on all the equipment in the cell. Hence work passes through the hands of several different cell workers and there may be an inventory of partially completed jobs held within the cell. Since a supervisor attending the cell is able to see which machines and which operators have open time, he or she can apply judgment in the ongoing assignment of available work.

Another scheduling discipline is to allow the cell workers to decide the sequence of dispatching jobs to the cell. *Worker dispatching* is sometimes used in facilities where workers are able to operate all or most machines in the cell. Production is planned by production management. The raw materials to be machined for the next few days of production are placed in storage at the beginning of the GT cell or flow line along with the routing cards. One worker may take a job through all machining operations. As the worker finishes one job, he or she returns to the beginning of the line to select another. If conflicts arise due to machine availability, a worker may start another job while waiting for a machine to clear. The entire intracell scheduling process is controlled by the workers themselves; a reasonable level of cooperation is required for smooth operation.

8.3.3. SOME ISSUES IN THE IMPLEMENTATION OF GROUP TECHNOLOGY

The primary drawback in exchanging a traditional process layout for group technology is the loss of flexibility. Since GT cells are organized around a specific group of components, machines are no longer interchangeable. Thus a reasonably stable product mix is required to ensure an economically viable degree of cell utilization.

A primary benefit of group technology implementation is the foundation it establishes for the application of automation. Group production reduces the variability in component geometry that is routed through any particular set of machines. This may reduce the materials-handling problem to a point where reprogrammable automation, such as robots, can be utilized in transferring workpieces between machines. The simplification of the production system made possible by employing group technology principles combined with reduced materials handling difficulties has been the basis around which many commercially successful automated machining center concepts and flexible manufacturing systems (FMS) have been developed. These automated production systems are the topic of the next section.

8.4. Flexible Manufacturing Systems: An Introduction

As presented earlier in Chapter 1 and in the beginning of this chapter, there are two major types of production layouts: (1) flow line (for mass production), which has received a great deal of attention in both research and implementation areas; and (2) process layout (for job shop or batch manufacturing), where 75% of the dollar volume of metalworked products are manufactured in batches of fewer than 50 parts [7]. The introduction of computers has made it easier to attack the complicated problems related to batch manufacturing. The concept of a flexible manufacturing system (FMS) is an attempt to apply computer controls to production scheduling, the control of machines, and the movement of materials in a discrete parts manufacturing environment.

8.4.1. THE FMS CONCEPT

We define the *flexible manufacturing system* as: general-purpose manufacturing machines, which are quite versatile and capable of performing different types of operations, linked together by materials-handling systems. Both the manufacturing machines and the materials-handling systems are under the control of a central computer system. There are two main objectives of employing FMS: (1) to provide full capability for random order production, permitting the machining of any desired mix of parts in a given period of time; and (2) to reduce the work-in-process and increase machine utilization in small-lot manufacturing.

It has been shown that in the conventional small batch production systems, as the part goes from raw stock to finished part by machining, it spends only about 5% of its time on a machine tool, where it is being set up and machined. But even there, it is actually being machined less than 30% of the time, resulting in useful work being done on the part less than $1\frac{1}{2}\%$ of the time the part takes to be processed through the shop. It is such factors that account to a large degree for the great amount of work-in-process and the low utilization of machines in small-lot production [18].

In FMS, machine tools are linked together by conveyor systems with relatively small buffers; work-in-process is not allowed to accumulate. Machine tools are equipped with multiple toolholders in order to minimize setup time between machining different components, thus minimizing total time in the manufacturing facility. The scheduling of components through the system to avoid excessive queueing while maintaining high machine utilization is a critical component to the overall success of the system operation.

The concept of FMS has been extended to the development of manufacturing cells where a group of four or five NC (*numerically controlled*) machine tools are arranged in a circle around a single robot, which does all the part handling and machine loading and unloading in the cell. These cells are typically designed to handle a family or group of parts and are usually limited to the machining of that group. Hence these new *machine center* concepts represent the marriage of programmable automation with group technology production organization principles.

One of the most important development areas in flexible manufacturing is in machine changeover. Machines within an FMS cell are equipped with tooling storage magazines which contain all the tooling the machine requires to perform work on the family of parts flowing through the cell. As parts flow through the FMS, these tools are changed automatically with a minimum of setup. As we have pointed out in the case of Kanban and GT, this naturally leads to the consideration of smaller lot sizes. In effect, the direction of research in scheduling FMS is toward scheduling on the basis of units, as opposed to lots. The very low setup times experienced in FMS allows unit production to approach the cost of batch production, which has the secondary effect of reducing the amount of finished component inventory to be carried.

8.4.2. ISSUES IN FMS

There are many problem areas in the design and selection of flexible manufacturing systems that are still open issues. The following partial list suggests some useful areas of research.

1. Flexible manufacturing systems are large scale manufacturing systems in terms of the number of work stations that are directly linked through automated materials handling systems. The small scale optimization models for job sequencing and routing, presented in Chap. 7, do not apply as well to FMS as to conventional manufacture. New analytical models must be developed or approximations for large scale systems introduced.
2. The advent of unmanned machining systems, i.e. an FMS with sufficient storage that it can operate for a considerable period without attention, has revealed that some realistic models ought to be developed to determine the size and location of storage that need to be considered.
3. Some modeling as analytical procedures that address the question of when to use dedicated transfer lines as opposite to FMS would be useful.
4. FMS demands much greater integration of the various manufacturing functions. For example, production, inspection, and design can be integrated in the total production task. Ideally, the inspection of a part should occur while the tool is cutting, and measurement results at the

point of cutting should be fed back to the machine controls with automatic adjustment of tool-tip position or machine speed, adaptive control, or in-process control. The question of measuring during the machining cycle is extremely complex but has considerable bearing on the feasibility of integrated inspection on an automatic production line. Techniques for quality control and nonconformance should be developed for FMS.

5. The design of manufacturing cells requires analytical models to determine the optimal number of machines to be assigned to the robot, taking into consideration the failure and repair rates of the robot.

8.4.3. FMS AND COMPUTER-INTEGRATED MANUFACTURING SYSTEMS

To date, there have been over 100 systems implemented under the FMS concept. The vast majority of these systems exist as islands of automation in which families of products have been selected for FMS manufacture, while conventional techniques of production are applied to the remainder. Materials handling to and from the FMS may be by forklift truck or other manned systems.

FMS has provided a foundation for the general concept of factory automation in discrete-parts manufacturing. In this concept, a programmable materials-handling system ties FMS cells together and connects them to automated warehouses. A connecting system might be a programmable automated guided vehicle system and an automated warehouse might consist of several automated storage and retrieval systems. Such computer-controlled materials-handling systems offer the potential for full factory automation.

Figure 8-7 shows a conceptual layout of a fully automated factory with both main and mini automated storage retrieval systems. The automated materials-handling system is considered a major component of factory automation. It is reported that excellent opportunities for significant gains (or cost reduction) lie in the area of materials handling. It must be made clear that the full automation of manufacturing processes without the coincidental automation of an integrated materials-handling system results in a nonjustifiable expenditure of capital [24].

As we mentioned earlier, one of the main objectives of FMS is to reduce the work-in-process. To assist in this objective, computer-controlled materials-handling systems are needed to ensure the arrival of material to workstations just when it is needed (not before or after) and to transport the products away from the workstation when the work on these products is completed at the workstation. Two materials-handling systems that are important in performing these functions are the automated storage and retrieval system and the automated guided vehicle system.

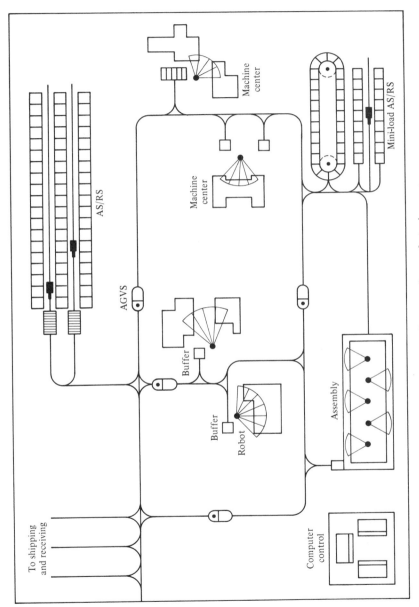

Figure 8-7. Computer-integrated manufacturing system.

Automated Storage/Retrieval Systems (AS/RS). A typical AS/RS consists of multiple storage racks with aisles between each pair of storage racks to allow for a moving crane (or moving vehicle) which is used for storage and retrieval of orders from the racks. The crane or vehicle is computer controlled. These automated storage/retrieval systems are used for two purposes:

1. To store the incoming raw material, in-process products and final assembly
2. To retrieve material needed in the manufacturing process and fill orders to service customer demand

In both storage and retrieval functions, we are interested in determining the optimal schedules which minimize the total distance (time or cost) traveled in the warehouse. Criteria for assigning and locating items on the storage racks need to be developed. The cost of order picking or retrieval is dependent on the structure of the orders to be retrieved, the capacity of the S/R crane, and the locations of the items in the warehouse. Wilson [25] states that order-picking costs typically dominate stock setting costs (cost of placing items in storage), and the former tend to be more sensitive to item locations than do the latter.

A typical large-scale AS/RS is designed to accommodate 20,000 storage locations at a load weight of 1500 lb each. The overall system dimensions are: 700 ft long, 100 ft wide, and 90 ft high. The horizontal and vertical speeds of the crane (S/R machine) are 500 and 90 ft/min, and the acceleration and deceleration are 1 ft/sec^2.

The mini-load AS/RS is usually installed inside the manufacturing system as shown in Fig. 8-7. This mini-load system is used as a temporary storage space for parts and components that are soon to be needed for assembly and manufacturing processes. The capacity of such a system is less than 500 lb and it is less than 30 ft long. Raw materials are not delivered directly to this mini-load system, but finished products could be temporarily stored until they are either shipped to other locations (or distribution centers) or transferred to the main AS/RS.

The main production control issues that are to be studied when designing AS/RS can be summarized as follows:

1. Approaches for optimal retrieval and storage of items. As orders are received for storage or retrieval, decisions must be made to group some orders and retrieve or store in the same tour such that the total time needed to perform these functions is minimized. For example, if the S/R machine is to move from location (x_1, y_1) to location (x_2, y_2) with speeds v_1 and v_2 in the horizontal and vertical directions respectively, then the time required to reach the new location is t_1 and is given by:

$$t_1 = \min \left\{ \frac{|x_1 - x_2|}{v_1}, \frac{|y_1 - y_2|}{v_2} \right\} \tag{8.9}$$

Also, constraints on the loading capacity of the S/R machine should be considered in the analysis as well as priority of orders and the frequency of the use of a product.

2. The warehouse configuration is dependent on the structure and the pattern of orders. Therefore, models must be developed to relate the order structure with the warehouse configuration. For example, order patterns that follow a given probability distribution may be stored and retrieved by using a specific algorithm for storage and retrieval of material. This, in turn, will affect the location of materials in the warehouse as well as their space requirements.

3. The physical structure of the storage racks will also depend on many factors, such as the size of the items (products) and the loading capacity of the S/R machine. Optimization models are needed to compare the performance of single-deep rack versus double-deep rack and single-wide aisle versus double-wide aisle.

Automatic Guided Vehicle Systems (AGVS). An automated guided vehicle system is defined as a system that includes vehicles, guidepath, controls, and interface for automatic routing and positioning. Most AGVs rely on conventional lead-acid batteries for power. Also, both optical and magnetic guidance techniques are used for AGVs. For optical guidance, vehicles track a guidepath placed on top of the floor. A high contrast between the floor and the guidepath is required for reliable tracking. A reflective tape or painted stripe on the floor is a common method. The vehicle focuses a light beam on the guidepath and by measuring the amplitude of the reflective light is able to track the path accurately. A unique ultraviolet system is used in another application. A nearly invisible chemical path is painted on the floor surface and the vehicle directs an ultraviolet light on the path. The chemical guidepath responds at a different wavelength for vehicle guidance [1].

For magnetic guidance, a slot is made into the floor surface and one or more wires are placed in the floor. Once the wires are placed in the floor, the slot is filled with a grout or epoxy material for a smooth, level finish. The wires are energized with a low-current, low-frequency ac signal which generates a magnetic field around the wire. Vehicle sensors detect this magnetic field for accurate tracking [1].

There are several objectives for the use of AGVs:

1. AGVs can be used as unit load transporters to carry individual loads on the deck of the vehicle. The typical loading capacity is in the range 4000 to 8000 lb, while the typical speed is 200 to 260 ft/min. The AGVs can also be used for feeding line stock to manufacturing departments.

2. AGVs are used as *pallet trucks* or *stop-and-drop* vehicles with the ability to carry up to two loads at a time.

3. The AGVs can interface with a work-in-process AS/RS storage system and the user's shop floor control system. The computer-controlled

AGVs provide the link for real-time material tracking throughout the facility. In addition, floor stock can be reduced significantly due to rapid, accurate delivery of material from central storage.

4. Guided vehicles can be used as mobile work platforms to provide the flexibility needed in assembly operations, particularly when a large variation in models is required. Line balancing can be simplified by using the parallel routing capability of guided vehicles.

Designing an automated guided vehicle system is a complex task. Besides hardware considerations, the design engineer should assess the impact on facilities layout, material procurement policy, and production policy [16]. The first step in the design is to determine:

1. Activity rates, such as loading and unloading, assembly times, and machining times for systems that interact with the AGVs. Both the average and peak loads must be determined.
2. The minimum and maximum sizes and weights of loads to be transferred.
3. Facility layout and restrictions, such as ramps, aisle widths, and columns.
4. The control requirements of the AGVs with the mini-load AS/RS, central AS/RS, manufacturing cells, and so on.

In addition to the hardware design, the control and functional requirements should be thoroughly analyzed; for example, the following questions should be addressed:

1. What are the operating rules of the AGVs?
2. How do the AGVs relate to other systems in the facility?
3. What determines the material movement, and how are destinations assigned?
4. Are priorities required, and how are those priorities assigned?
5. How will the aisles be used?
6. What will the direction of movement be in each aisle?
7. Should movement be permitted in both directions?
8. Should there be bypasses at pickup and drop-off points? If so, how long should they be?

Maxwell and Muckstadt [16] presented some analysis techniques and methodologies to determine the minimum number of vehicles required as well as the vehicle routes and the number of trips over each route such that the movement requirements are met.

8.4.4. PRODUCTION CONTROL IN THE AUTOMATED FACTORY

As the reader should appreciate by now, automated factory concepts will add a new dimension and a new level of complexity to the production control problem. In a less automated factory, variations from plan arising from uncertainty in the production process could be adjusted by the intelligent intervention of a worker or shop floor supervisor. Thus, if a machine breaks down or if material shortages occur due to faulty inventory counts, adjustments and/or substitutions would be made by those people on the spot. In fact, adjusting plans daily is a way of life in many factory environments. True automation implies that the planning and the adjustment process will be relegated to the computer and performed automatically.

To the extent that this will occur, the production control process will have to become considerably more precise than it currently is in the average manufacturing firm. Through experience in the use of computerized MRP systems, American manufacturers have become acutely conscious of the importance of accurate data collection in any computerized decision-making process. Thus, in the automated factory, computerized data collection systems, which will become the cornerstone of decision making, will have to be exact and complete. Currently, automated data collection on the tool-life and other performance characteristics of *computer numerically controlled* (CNC) machines is available. Similarly, material counts in AS/RS systems are kept track of by the computer that directs and controls storage and retrieval.

Computerized decision making raises some new issues in the location of decision making; in particular, which decisions should be central decisions and which decisions should be localized? In the less automated production processes of today, it is typical for an MRP system to generate weekly or biweekly work-center loads. This is done in the interest of providing some coordination in the completion times of related parts. The overall loading of the shop is done centrally. However, individual work centers may be provided with priority indicators by job but allowed to use their own disgression in the scheduling of individual jobs. This is a sensible approach because the shop floor supervisor, who assigns jobs to workers and machines, is familiar with the skills of individual workers and can make better assignments than could be expected by some simple scheduling rule. Under automation, without workers in the system and with complete information on machine capabilities, there is no reason to believe that local decisions cannot be centralized or at least subject to the real-time intervention of a central computer which is responsible for overall process control.

Closely related to the question of local versus central decision making is the issue of local versus global process optimization. In the case of the shop floor supervisor described previously, who assigns jobs to worker/machine combinations, decisions are made that may lead to more optimal local performance. Hopefully, in so doing, the decision of the shop floor supervisor may very well be consistent with better overall or global performance. However, there is the question of the effect of these decisions on related departments. For example, the shop floor supervisor may accelerate jobs that are related to other jobs being delayed in another department. The relationships, for example, might be that both jobs are needed at a particular point for a common assembly. When one job arrives early, it results in unwanted work-in-process. In this case, the uncoordinated local optimal decisions are not leading in the direction of global optimality.

In the automated factory it will be possible at least to know, in real time, the relationship of different decisions that affect one another by the continuous computerized data collection systems. However, having the information does not necessarily mean that some form of optimization can be employed globally. As the reader should appreciate from Chapter 7, scheduling algorithms become very complex as soon as one is dealing with more than a few machines. The optimization of an entire factory of machining centers, respecting the coordination of all activities, is a problem of much more immense proportion. However, one can say that better (more optimal) decisions should be possible within the context of a real-time information system that has the capability of continuously evaluating data. The algorithms that will have to be developed to do all this or the implementation and coordination of existing algorithms are still a matter of considerable future research and development.

It is probably safe to say that in most manufacturing firms, total computerization of the production control process would be economically difficult given today's level of uncertainty in the production process. However, a large portion of that uncertainty is related to variability in human performance, which is eliminated with automation. To the extent that uncertainty is removed from within the manufacturing process itself, the mathematical tools of production control that are required will be less complex than they otherwise would have to be. With the limitations in today's programmable automation machinery, the economically effective application of total automation in any production system will depend to a large extent on how much simplification can be designed into the production process (e.g., using group technology) and how much uncertainty can be engineered out of that process.

REFERENCES

[1] ADAMS, WALT, "AGVS Horizontal Transportation Systems," *Advanced Institute of Materials Handling Teachers*, Auburn University, Alabama, (June 12–17) 1983.

[2] BOUCHER, THOMAS O., "Lot Sizing in Group Technology Production Systems," *International Journal of Production Research*, Vol. 22, no. 1 (1984).

[3] BOUCHER, THOMAS O., AND JOHN A. MUCKSTADT, "Cost Estimating Methods for Evaluating the Conversion From a Functional Manufacturing Layout to Group Technology," Technical Report no. 603, Cornell University, Ithaca, New York, 1983.

[4] BURBIDGE, JOHN L., "Production Flow Analysis," *The Production Engineer*, April/May, 1971.

[5] BUZACOTT, JOHN A. AND J. G. SHANTHIKUMAR, "Models for Understanding Flexible Manufacturing Systems," *AIIE Transactions*, Vol. 12, No. 4, Dec. 1980.

[6] BUZACOTT, JOHN A. AND DAVID D. W. YAO, "Flexible Manufacturing Systems: A Review of Models," Working paper no. 82-007, Dept. of Industrial Engineering, University of Toronto, Ontario.

[7] COOK, N. H., "Computer-Managed Parts Manufacture," *Scientific American*, 22, no. 2 (1975) 232.

[8] DEVRIES, MARTIN F., SUSAN M. HARVEY, AND VIJAY A. TIPNIS, *Group Technology, an Overview and Bibliography*, Publication No. MDC 76-601, Army Materials and Mechanics Research Center, August, 1976.

[9] ELSAYED, ELSAYED A., AND RICHARD G. STERN, "Computerized Algorithms for Order Processing in Automated Warehousing Systems," *International Journal of Production Research*, Vol. 21, No. 4, July/Aug. 1983.

[10] FOO, F. C. AND J. G. WAGNER, "Setup Times in Cyclic and Acyclic Group Technology Scheduling Systems," *International Journal of Production Research*, Vol. 21, No. 1, Jan./Feb. 1983.

[11] *The FMS Magazine*, "Big System at John Deere Solves New Product Problems," 1, no. 1 (1982) pp. 16–19.

[12] GALLAGHER, C. C. AND W. A. KNIGHT, *Group Technology*. London, England: Butterworths, 1973.

[13] GODDARD, WALTER E., "Kanban Versus MRP II—Which is Best For You?", *Modern Materials Handling*, November, 1982.

[14] GUPTA, RAJIV M. AND JAMES A. TOMPKINS, "An Examination of the Dynamic Behavior of Part-Families in Group Technology," *International Journal of Production Research*, Vol. 20, No. 1, Jan./Feb. 1982.

[15] HITOMI, KATSUNDO, AND INYONG HAM, "Machine Loading for Group Technology Applications," *Annals of CIRP*, 25, 1977.

[16] MAXWELL, WILLIAM L., AND JACK A. MUCKSTADT, "Design of Automatic Guided Vehicle System," Technical Report no. 504, School of Operations Research and Industrial Engineering, Cornell University, 1981.

[17] MERCHANT, M. EUGENE, "The Factory of the Future—Technological Aspects," *Towards the Factory of the Future*, PED 1, (1980) pp. 77–82.

[18] MERCHANT, M. EUGENE, "World Trends in Flexible Manufacturing Systems", *The FMS Magazine*, 1, no. 1 (1982) pp. 4–5.

[19] MONDEN, YASUHIRO, "The Toyota Production System," *Industrial Engineering*, January, 1981.

[20] NAKAMURA, NOBUTO, TERIRHIKO YOSHIDA, AND KATSUNDO HITOMI, "Group Production Scheduling for Minimum Total Tardiness," *AIIE Transactions*, Vol. 10, No. 2, June 1978.

[21] ORLICKY, JOSEPH A., *Material Requirements Planning*. New York, N.Y.: McGraw-Hill Book Co., 1975.

[22] ROSENBLATT, MEIR J. AND NACHUM FINGER, "An Application of Grouping Procedure to a Multi-Item Production System," *International Journal of Production Research*, Vol. 21, No. 2, Mar./Apr. 1983.

[23] SEIDEL, RAINER, "Optimization of the Availability of Complex Manufacturing Systems—Methods and Examples," *International Journal of Production Research*, Vol. 21, No. 2, Mar./Apr. 1983.

[24] TOMPKINS, JAMES A., "The Automated Factory," *National Material Handling Forum*, April 1983.

[25] WILSON, H. G., "Order Quantity, Product Popularity, and the Location of Stock in a Warehouse," *AIIE Transactions*, 9, no. (1977), p. 230.

APPENDIX A

Proof of Equation 3.65

For the normal distribution,

$$\bar{S}(I_{max}, T) = \int_{I_{max}}^{\infty} (x - I_{max})g(x, l + T) \, dx$$

$$= \sigma\phi(z) - [(I_{max} - \mu)(1 - \Phi(z))] \tag{3.65}$$

Proof

$$\bar{S}(I_{max}, T) = \int_{I_{max}}^{\infty} (x - I_{max}) \frac{1}{\sqrt{2\pi}\sigma} \exp\left[-\frac{1}{2}\left(\frac{x - \mu}{\sigma}\right)^2\right] dx$$

Let $z = (x - \mu)/\sigma$, $x = \mu + z\sigma$; then

$$\bar{S}(I_{max}, T) = \int_{(I_{max} - \mu)/\sigma}^{\infty} (\mu + \sigma z - I_{max})\phi(z) \, dz$$

$$= \sigma \int_{(I_{max} - \mu)/\sigma}^{\infty} z\phi(z) \, dz + (\mu - I_{max}) \int_{(I_{max} - \mu)/\sigma}^{\infty} \phi(z) \, dz$$

Evaluating the first term yields

$$\int_{(I_{max}-\mu)/\sigma}^{\infty} z\phi(z)\ dz = \frac{1}{\sqrt{2\pi}} \int_{(I_{max}-\mu)/\sigma}^{\infty} z e^{-z^2/2}\ dz$$

Let $v = z^2/2$, $dv = z\ dz$; then

$$\frac{1}{\sqrt{2\pi}} \int_{(I_{max}-\mu)/\sigma}^{\infty} z e^{-(z^2/2)}\ dz = \frac{1}{\sqrt{2\pi}} \int_{(1/2)[(I_{max}-\mu)/\sigma]^2}^{\infty} e^{-v}\ dv$$

$$= -\frac{1}{\sqrt{2\pi}} e^{-v}\ \Big|_{(1/2)[(I_{max}-\mu)/\sigma]^2}^{\infty}$$

$$= \frac{1}{\sqrt{2\pi}} \exp\left[-\frac{1}{2}\left(\frac{I_{max}-\mu}{\sigma}\right)^2\right] = \phi(z)$$

Evaluating the second term yields

$$(\mu - I_{max}) \int_{(I_{max}-\mu)/\sigma}^{\infty} \phi(z)\ dz = (\mu - I_{max})[1 - \Phi(z)]$$

Therefore,

$$\bar{S}(I_{max},\ T) = \sigma\phi(z) - (I_{max} - \mu)[1 - \Phi(z)]$$

APPENDIX B

THE CUMULATIVE STANDARDIZED NORMAL DISTRIBUTION FUNCTION
(Note: $.0^{2}1350 = .001350$)

Entry $= P\{Z < Z_{\alpha}\} = \alpha$

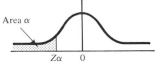

Area α

$Z\alpha$ 0

Z_{α}	.00	.01	.02	.03	.04	.05	.06	.07	.08	.09
$-.0$	.5000	.4960	.4920	.4880	.4840	.4801	.4761	.4721	.4681	.4641
$-.1$	.4602	.4562	.4522	.4483	.4443	.4404	.4364	.4325	.4286	.4247
$-.2$	.4207	.4168	.4219	.4090	.4052	.4013	.3974	.3936	.3897	.3859
$-.3$	.3821	.3783	.3745	.3707	.3669	.3632	.3594	.3557	.3520	.3483
$-.4$	.3446	.3409	.3372	.3336	.3300	.3264	.3228	.3192	.3156	.3121
$-.5$	.3085	.3050	.3015	.2981	.2946	.2912	.2877	.2843	.2810	.2776
$-.6$	.2743	.2709	.2676	.2643	.2611	.2578	.2546	.2514	.2483	.2451
$-.7$	.2420	.2389	.2358	.2327	.2297	.2266	.2236	.2206	.2177	.2148
$-.8$	.2119	.2090	.2061	.2033	.2005	.1977	.1949	.1922	.1894	.1867
$-.9$	.1841	.1814	.1788	.1762	.1736	.1711	.1685	.1660	.1635	.1611
-1.0	.1587	.1562	.1539	.1515	.1492	.1469	.1446	.1423	.1401	.1379
-1.1	.1357	.1335	.1314	.1292	.1271	.1251	.1230	.1210	.1190	.1170
-1.2	.1151	.1131	.1112	.1093	.1075	.1056	.1038	.1020	.1003	.09853
-1.3	.09680	.09510	.09342	.09176	.09012	.08851	.08691	.08534	.08379	.08226
-1.4	.08076	.07927	.07780	.07636	.07493	.07353	.07215	.07078	.06944	.06811
-1.5	.06681	.06552	.06426	.06301	.06178	.06057	.05938	.05821	.05705	.05592
-1.6	.05480	.05370	.05262	.05155	.05050	.04947	.04846	.04746	.04648	.04551
-1.7	.04457	.04363	.04272	.04182	.04093	.04006	.03920	.03836	.03754	.03673
-1.8	.03593	.03515	.03438	.03362	.03288	.03216	.03144	.03074	.03005	.02938
-1.9	.02872	.02807	.02743	.02680	.02619	.02559	.02500	.02442	.02385	.02330
-2.0	.02275	.02222	.02169	.02118	.02068	.02018	.01970	.01923	.01876	.01831
-2.1	.01786	.01743	.01700	.01659	.01618	.01578	.01539	.01500	.01463	.01426
-2.2	.01390	.01355	.01321	.01287	.01255	.01222	.01191	.01160	.01130	.01101
-2.3	.01072	.01044	.01017	$.0^{2}9903$	$.0^{2}9642$	$.0^{2}9387$	$.0^{2}9137$	$.0^{2}8894$	$.0^{2}8656$	$.0^{2}8424$
-2.4	$.0^{2}8198$	$.0^{2}7976$	$.0^{2}7760$	$.0^{2}7549$	$.0^{2}7344$	$.0^{2}7143$	$.0^{2}6947$	$.0^{2}6756$	$.0^{2}6569$	$.0^{2}6387$
-2.5	$.0^{2}6210$	$.0^{2}6037$	$.0^{2}5868$	$.0^{2}5703$	$.0^{2}5543$	$.0^{2}5386$	$.0^{2}5234$	$.0^{2}5085$	$.0^{2}4940$	$.0^{2}4799$
-2.6	$.0^{2}4661$	$.0^{2}4527$	$.0^{2}4396$	$.0^{2}4269$	$.0^{2}4145$	$.0^{2}4025$	$.0^{2}3907$	$.0^{2}3793$	$.0^{2}3681$	$.0^{2}3573$
-2.7	$.0^{2}3467$	$.0^{2}3364$	$.0^{2}3264$	$.0^{2}3167$	$.0^{2}3072$	$.0^{2}2980$	$.0^{2}2890$	$.0^{2}2803$	$.0^{2}2718$	$.0^{2}2635$
-2.8	$.0^{2}2555$	$.0^{2}2477$	$.0^{2}2401$	$.0^{2}2327$	$.0^{2}2256$	$.0^{2}2186$	$.0^{2}2118$	$.0^{2}2052$	$.0^{2}1988$	$.0^{2}1926$
-2.9	$.0^{2}1866$	$.0^{2}1807$	$.0^{2}1750$	$.0^{2}1695$	$.0^{2}1641$	$.0^{2}1589$	$.0^{2}1538$	$.0^{2}1489$	$.0^{2}1441$	$.0^{2}1395$
-3.0	$.0^{2}1350$	$.0^{2}1306$	$.0^{2}1264$	$.0^{2}1223$	$.0^{2}1183$	$.0^{2}1144$	$.0^{2}1107$	$.0^{2}1070$	$.0^{2}1035$	$.0^{2}1001$

THE CUMULATIVE STANDARDIZED NORMAL DISTRIBUTION FUNCTION
(Note: $.9^{2}8650 = .998650$)

Entry $= P\{Z < Z_{1-\alpha}\} = 1 - \alpha$

Area $1 - \alpha$

$0 \quad Z_{1-\alpha}$

$Z_{1-\alpha}$	.00	.01	.02	.03	.04	.05	.06	.07	.08	.09
.0	.5000	.5040	.5080	.5120	.5160	.5199	.5239	.5279	.5319	.5359
.1	.5398	.5438	.5478	.5517	.5557	.5596	.5636	.5675	.5714	.5753
.2	.5793	.5832	.5871	.5910	.5948	.5987	.6026	.6064	.6103	.6141
.3	.6179	.6217	.6255	.6293	.6331	.6368	.6406	.6443	.6480	.6517
.4	.6554	.6591	.6628	.6664	.6700	.6736	.6772	.6808	.6844	.6879
.5	.6915	.6950	.6985	.7019	.7054	.7088	.7123	.7157	.7190	.7224
.6	.7257	.7291	.7324	.7357	.7389	.7422	.7454	.7486	.7517	.7549
.7	.7580	.7611	.7642	.7673	.7703	.7734	.7764	.7794	.7823	.7852
.8	.7881	.7910	.7939	.7967	.7995	.8023	.8051	.8078	.8106	.8133
.9	.8159	.8186	.8212	.8238	.8264	.8289	.8315	.8340	.8365	.8389
1.0	.8413	.8438	.8461	.8485	.8508	.8531	.8554	.8577	.8599	.8661
1.1	.8643	.8665	.8686	.8708	.8729	.8749	.8770	.8790	.8810	.8830
1.2	.8849	.8869	.8888	.8907	.8925	.8944	.8962	.8980	.8997	.90147
1.3	.90320	.90490	.90658	.90824	.90988	.91149	.91309	.91466	.91621	.91774
1.4	.91924	.92073	.92220	.92364	.92507	.92647	.92785	.92922	.93056	.93189
1.5	.93319	.93448	.93574	.93699	.93822	.93943	.94062	.94179	.94295	.94408
1.6	.94520	.94630	.94738	.94845	.94950	.95053	.95154	.95254	.95352	.95449
1.7	.95543	.95637	.95728	.95818	.95907	.95994	.96080	.96164	.96246	.96327
1.8	.96407	.96485	.96562	.96638	.96712	.96784	.96856	.96926	.96995	.97062
1.9	.97128	.97193	.97257	.97320	.97381	.97441	.97500	.97558	.97615	.97670
2.0	.97725	.97778	.97831	.97882	.97932	.97982	.97030	.98077	.98124	.98169
2.1	.98214	.98257	.98300	.98341	.98382	.98422	.98461	.98500	.98537	.98574
2.2	.98610	.98645	.98679	.98713	.98745	.98778	.98809	.98840	.98870	.98899
2.3	.98928	.98956	.98983	$.9^{2}0097$	$.9^{2}0358$	$.9^{2}0613$	$.9^{2}0863$	$.9^{2}1106$	$.9^{2}1344$	$.9^{2}1576$
2.4	$.9^{2}1802$	$.9^{2}2024$	$.9^{2}2240$	$.9^{2}2451$	$.9^{2}2656$	$.9^{2}2857$	$.9^{2}3053$	$.9^{2}3244$	$.9^{2}3431$	$.9^{2}3613$
2.5	$.9^{2}3790$	$.9^{2}3963$	$.9^{2}4132$	$.9^{2}4297$	$.9^{2}4457$	$.9^{2}4614$	$.9^{2}4766$	$.9^{2}4915$	$.9^{2}5060$	$.9^{2}5201$
2.6	$.9^{2}5339$	$.9^{2}5473$	$.9^{2}5604$	$.9^{2}5731$	$.9^{2}5855$	$.9^{2}5975$	$.9^{2}6093$	$.9^{2}6207$	$.9^{2}6319$	$.9^{2}6427$
2.7	$.9^{2}6533$	$.9^{2}6636$	$.9^{2}6736$	$.9^{2}6833$	$.9^{2}6928$	$.9^{2}7020$	$.9^{2}7110$	$.9^{2}7197$	$.9^{2}7282$	$.9^{2}7365$
2.8	$.9^{2}7445$	$.9^{2}7523$	$.9^{2}7599$	$.9^{2}7673$	$.9^{2}7744$	$.9^{2}7814$	$.9^{2}7882$	$.9^{2}7948$	$.9^{2}8012$	$.9^{2}8074$
2.9	$.9^{2}8134$	$.9^{2}8193$	$.9^{2}8250$	$.9^{2}8305$	$.9^{2}8359$	$.9^{2}8411$	$.9^{2}8462$	$.9^{2}8511$	$.9^{2}8559$	$.9^{2}8605$
−3.0	$.9^{2}8650$	$.9^{2}8684$	$.9^{2}8736$	$.9^{2}8777$	$.9^{2}8817$	$.9^{2}8856$	$.9^{2}8893$	$.9^{2}8930$	$.9^{2}8965$	$.9^{2}8999$

From Isaac N. Gibra, *Probability and Statistical Inference for Scientists and Engineers*, © 1973, p. 562. Reprinted by permission of Prentice-Hall, Inc., Englewood Cliffs, N.J.

PERCENTAGE POINTS OF THE *t* DISTRIBUTION

v	0.45	0.40	0.35	0.30	0.25	α 0.125	0.05	0.025	0.0125	0.005	0.0025
1	0.158	0.325	0.510	0.727	1.000	2.414	6.314	12.71	25.45	63.66	127.3
2	0.142	0.289	0.445	0.617	0.817	1.604	2.920	4.303	6.205	9.925	14.09
3	0.137	0.277	0.424	0.584	0.765	1.423	2.353	3.183	4.177	5.841	7.453
4	0.134	0.271	0.414	0.569	0.741	1.344	2.132	2.776	3.495	4.604	5.598
5	0.132	0.267	0.408	0.559	0.727	1.301	2.015	2.571	3.163	4.032	4.773
6	0.131	0.265	0.404	0.553	0.718	1.273	1.943	2.447	2.969	3.707	4.317
7	0.130	0.263	0.402	0.549	0.711	1.254	1.895	2.365	2.841	3.500	4.029
8	0.130	0.262	0.399	0.546	0.706	1.240	1.860	2.306	2.752	3.355	3.833
9	0.129	0.261	0.398	0.543	0.703	1.230	1.833	2.262	2.685	3.250	3.690
10	0.129	0.260	0.397	0.542	0.700	1.221	1.813	2.228	2.634	3.169	3.581
11	0.129	0.260	0.396	0.540	0.697	1.215	1.796	2.201	2.593	3.106	3.500
12	0.128	0.259	0.395	0.539	0.695	1.209	1.782	2.179	2.560	3.055	3.428
13	0.128	0.259	0.394	0.538	0.694	1.204	1.771	2.160	2.533	3.012	3.373
14	0.128	0.258	0.393	0.537	0.692	1.200	1.761	2.145	2.510	2.977	3.326
15	0.128	0.258	0.393	0.536	0.691	1.197	1.753	2.132	2.490	2.947	3.286
20	0.127	0.257	0.391	0.533	0.687	1.185	1.725	2.086	2.423	2.845	3.153
25	0.127	0.256	0.390	0.531	0.684	1.178	1.708	2.060	2.385	2.787	3.078
30	0.127	0.256	0.389	0.530	0.683	1.173	1.697	2.042	2.360	2.750	3.030
40	0.126	0.255	0.388	0.529	0.681	1.167	1.684	2.021	2.329	2.705	2.971
60	0.126	0.254	0.387	0.527	0.679	1.162	1.671	2.000	2.299	2.660	2.915
120	0.126	0.254	0.386	0.526	0.677	1.156	1.658	1.980	2.270	2.617	2.860
∞	0.126	0.253	0.385	0.524	0.674	1.150	1.645	1.960	2.241	2.576	2.807

v = degrees of freedom
Reprinted with permission from *Probability and Statistics in Engineering and Management*, William W. Hines and Douglas C. Montgomery, John Wiley & Sons, 1972.

INDEX